DIGITAL ELECTRONICS

Reg 18 Voc Center
#10

DIGITAL
ELECTRONICS

ROGER L. TOKHEIM
HENRY SIBLEY HIGH SCHOOL

McGraw-Hill Book Company

Gregg Division

New York
St. Louis
Dallas
San Francisco
Auckland
Bogotá
Düsseldorf
Johannesburg
London
Madrid

Mexico
Montreal
New Delhi
Panama
Paris
São Paulo
Singapore
Sydney
Tokyo
Toronto

Library of Congress Cataloging in Publication Data

Tokheim, Roger.
   Digital electronics.

   Includes index.
   1. Digital electronics. I. Title.
TK7868.D5T65   621.3815   78-17949
ISBN 0-07-064954-5

Acknowledgments

Students, teachers, school administrators, and industrial trainers have contributed to the development of the *Basic Skills in Electricity and Electronics* series. Classroom testing of preliminary editions has been conducted at the following sites:

Burr D. Coe Vocational Technical High School (East Brunswick, New Jersey)
Chantilly Secondary School (Chantilly, Virginia)
Nashoba Valley Technical High School (Westford, Massachusetts)
Platt Regional Vocational Technical High School (Milford, Connecticut)
United States Steel Corporation: Edgar Thomson, Irvin Works (Dravosburg, Pennsylvania)

   The publisher gratefully acknowledges the helpful comments and suggestions received from these participants.

   34567890 WCWC   854321

The editors for this book were Gordon Rockmaker and Mark Haas. The designer was Tracy A. Glasner. The art supervisor was George T. Resch. Cover photography by Martin Bough/Studios Inc. The production supervisor was Kathleen Morrissey. It was set in Electra by Progressive Typographers.
Printed and bound by Webcrafters, Inc.

Contents

Editor's Foreword

The Gregg/McGraw-Hill *Basic Skills in Electricity and Electronics* series has been designed to provide entry-level competencies in a wide range of occupations in the electrical and electronics fields. The series consists of instructional materials geared especially for the career-oriented student. Each major subject area covered in the series is supported by a textbook, an activities manual, and a teacher's manual. All the materials focus on the theory, applications, and experiences required for those just beginning their chosen vocations.

There are two basic considerations in the preparation of educational materials for such a series: the needs of the learner and the needs of the employer. The materials in the series have been designed to meet these needs. They are based on many years of experience in the classroom and with electricity and electronics. In addition, these books reflect the needs of industry and commerce as developed through questionnaires, surveys, interviews with employers, government occupational trend reports, and various field studies.

Further refinements both in pedagogy and technical content resulted from actual classroom experience with the materials. Preliminary editions of selected texts and manuals were field-tested in schools and in-plant training programs throughout the country. The knowledge from this testing has enhanced the effectiveness and the validity of the materials.

Teachers will find the materials in each of the subject areas well coordinated and structured around a framework of modern objectives. Students will find the concepts clearly presented with many practical references and applications. In all, every effort has been made to prepare and refine the most effective learning tools possible.

The publisher and editor welcome comments from teachers and students using this book.

Charles A. Schuler
Project Editor

BASIC SKILLS IN ELECTRICITY AND ELECTRONICS

Charles A. Schuler, Project Editor

Books in this series

Introduction to Television Servicing by Wayne C. Brandenburg
Electricity: Principles and Applications by Richard J. Fowler
Instruments and Measurements by Charles M. Gilmore
Microprocessors by Charles M. Gilmore (*in preparation*)
Motors and Generators by Russell L. Heiserman and Jack D. Burson
Small Appliance Repair by Phyllis Palmore and Nevin E. André
Residential Wiring by Gordon Rockmaker
Electronics: Principles and Applications by Charles A. Schuler
Your Future in Electricity and Electronics by William A. Stanton
Digital Electronics by Roger L. Tokheim

Preface

Digital electronics is no longer a specialized field in electronics. It has become a subject that should be studied by anyone entering a career in electronics. First used in computing devices, digital circuits are now found in automobiles, communications devices, children's toys, and other consumer products. The microelectronic revolution has thus made the study of digital electronics a necessary part of a student's training in electronics.

The material in this text is based on carefully selected and formulated performance objectives. These objectives are accomplished through the use of digital systems and subsystems. The systems-subsystems approach is fundamental to digital electronics because of the extensive use of medium- and large-scale integrated circuits. By using small-scale integrated circuits, however, the basic fundamentals of *how* these systems and subsystems operate can be carefully developed. All circuits in the text can be wired for classroom demonstration. Each circuit has been tested by the author using low-cost TTL ICs.

Student motivation is one of the most difficult problems facing instructors. The text includes the following techniques which have been developed to maintain a high level of student motivation:

1. Simple, two-color illustrations emphasize important ideas and help students focus on basic concepts.
2. Frequent, short self tests (with answers) provide immediate reinforcement and build the student's confidence.
3. A unique format presents important terms and concepts in the margin of each page.
4. The systems-subsystems approach makes the topics more relevant to the student.
5. Simple analysis techniques build the student's ability to troubleshoot.
6. Practical, "hands-on" tasks are emphasized.

Students should have a general math background and an introduction to dc circuits before using the textbook. Binary mathematics and Boolean concepts are introduced and explained as needed. Digital electronics may be studied concurrently with a course in basic electronics since knowledge of active discrete components is not a prerequisite.

These materials reflect the experience and feedback of several years of classroom testing. The author welcomes comments and suggestions from users.

Roger L. Tokheim

Safety

Electric devices and circuits can be dangerous. Safe practices are necessary to prevent electrical shock, fires, explosions, mechanical damage, and injuries resulting from the improper use of tools.

Perhaps the greatest of these hazards is electrical shock. A current through the human body in excess of 10 milliamperes can paralyze the victim and make it impossible to let go of a "live" conductor. Ten milliamperes is a small amount of electrical flow: It is *ten one-thousandths* of an ampere. An ordinary flashlight uses more than 100 times that amount of current! If a shock victim is exposed to currents over 100 milliamperes, the shock is often *fatal*. This is still far less current than the flashlight uses.

A flashlight cell can deliver more than enough current to kill a human being. Yet it is safe to handle a flashlight cell because the resistance of human skin normally will be high enough to greatly limit the flow of electric current. Human skin usually has a resistance of several hundred thousand ohms. In low-voltage systems, a high resistance restricts current flow to very low values. Thus, there is little danger of an electrical shock.

High voltage, on the other hand, can force enough current through the skin to produce a shock. The danger of harmful shock increases as the voltage increases. Those who work on very high-voltage circuits must use special equipment and procedures for protection.

When human skin is moist or cut, its resistance can drop to several hundred ohms. Much less voltage is then required to produce a shock. Potentials as low as 40 volts can produce a fatal shock if the skin is broken! Although most technicians and electrical workers refer to 40 volts as a *low voltage*, it does not necessarily mean *safe voltage*. Obviously, you should, therefore, be very cautious even when working with so-called low voltages.

Safety is an attitude; safety is knowledge. Safe workers are not fooled by terms such as *low voltage*. They do not assume protective devices are working. They do not assume a circuit is off even though the switch is in the OFF position. They know that the switch could be defective.

As your knowledge of electricity and electronics grows, you will learn many specific safety rules and practices. In the meantime:

1. Investigate before you act
2. Follow procedures
3. When in doubt, *do not act:* ask your instructor

GENERAL SAFETY RULES FOR ELECTRICITY AND ELECTRONICS

Safe practices will protect you and those around you. Study the following rules. Discuss them with others. Ask your instructor about any that you do not understand.

1. Do not work when you are tired or taking medicine that makes you drowsy.
2. Do not work in poor light.
3. Do not work in damp areas.
4. Use approved tools, equipment, and protective devices.
5. Do not work if you or your clothing is wet.
6. Remove all rings, bracelets and similar metal items.
7. Never assume that a circuit is off. Check it with a device or piece of equipment that you are sure is operating properly.
8. Do not tamper with safety devices. *Never* defeat an interlock switch. Verify that all interlocks operate properly.
9. Keep your tools and equipment in good condition. Use the correct tool for the job.
10. Verify that capacitors have discharged. Some capacitors may store a lethal charge for a long time.

11. Do not remove equipment grounds. Verify that all grounds are intact.
12. Do not use adaptors that defeat ground connections.
13. Use only an approved fire extinguisher. Water can conduct electrical current and increase the hazards and damage. Carbon dioxide (CO_2) and certain halogenated extinguishers are preferred for most electrical fires. Foam types may also be used in some cases.
14. Follow directions when using solvents and other chemicals. They may explode, ignite, or damage electrical circuits.
15. Certain electronic components affect the safe performance of the equipment. Always use the correct replacement parts.
16. Use protective clothing and safety glasses when handling high-vacuum devices such as television picture tubes.
17. Do not attempt to work on complex equipment or circuits before you are ready. There may be hidden dangers.
18. Some of the best safety information for electrical and electronic equipment is in the literature prepared by the manufacturer. Find it and use it!

Any of the above rules could be expanded. As your study progresses, you will learn many of the details concerning proper procedure. Learn them well, because they are the most important information available.

Remember, always practice safety; your life depends on it.

Digital Electronics

Digital electronics is the world of the calculator, the computer, the integrated circuit, and the binary numbers 0 and 1. This is an exciting field within electronics because the uses for digital circuits are expanding so rapidly. One small integrated circuit can perform the task of thousands of transistors, diodes, and resistors. You see digital circuits in operation every day. At stores the cash registers read out digital displays. The tiny pocket calculators verge on becoming personal computers. All sizes of computers perform complicated tasks with fantastic speed and accuracy. Factory machines are controlled by digital circuits. Digital clocks and watches flash the time. Some automobiles use microprocessors to control several engine functions. Technicians use digital voltmeters and frequency counters.

All persons working in electronics must now understand digital electronic circuits. The inexpensive integrated circuit has made the subject of digital electronics easy to study. You will use many integrated circuits to construct digital circuits.

1-1 WHAT IS A DIGITAL CIRCUIT?

In your experience with electricity and electronics you have probably used analog circuits. The circuit in Fig. 1-1(a) puts out an *analog* signal or voltage. As the wiper on the potentiometer is moved upward, the voltage from points A to B *gradually* increases. When the potentiometer is moved downward, the voltage gradually decreases from 5 to 0 volts (V). The waveform diagram in Fig. 1-1(b) is a graph of the analog output. On the left side the voltage from A to B is gradually increasing to 5 V; on the right side the voltage is gradu-

ally decreasing to 0 V. By stopping the potentiometer wiper at any midpoint, we can get an output voltage anywhere between 0 and 5 V. An analog device, then, is one that has a signal which *varies continuously* in step with the input.

A digital device operates with a digital signal. Figure 1-2(a) pictures a square-wave generator. The generator produces a square waveform that is displayed on the oscilloscope. The digital signal is only at +5 V *or* at 0 V, as diagramed in Fig. 1-2(b). The voltage at point A moves from 0 to 5 V. The voltage then stays at a +5 V for a time. At point B the voltage drops immediately from +5 to 0 V. The voltage then stays at 0 V for a time. Only two voltages are present in a digital electronic circuit. In the waveform diagram in Fig. 1-2(b) these voltages are labeled

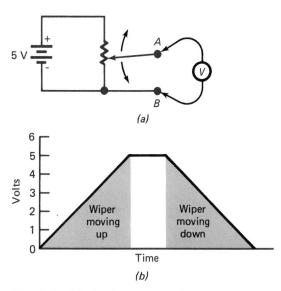

Fig. 1-1 (a) Analog output from a potentiometer. (b) Analog signal waveform.

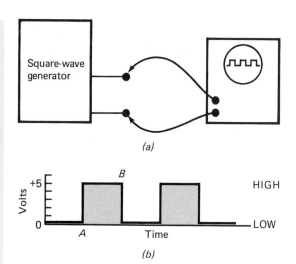

From page 1:
Analog signal

Digital signal

On this page:
Calculator

Digital computer

Microprocessor

Fig. 1-2 (*a*) Digital signal displayed on scope. (*b*) Digital signal waveform.

HIGH and LOW. The HIGH voltage is +5 V; the LOW voltage is 0 V. Later we shall call the HIGH voltage (+5 V) a logical 1 and the LOW voltage (0 V) a logical 0.

Circuits that handle only HIGH and LOW signals are called *digital circuits*. We mentioned that digital electronics is the world of logical 0s and 1s. The voltages in Fig. 1-2(*b*) are rather typical of the voltages you will be working with in digital electronics.

The digital signal in Figure 1-2(*b*) could also be generated by a simple on-off switch. A digital signal could also be generated by a transistor turning on and off. In recent years digital electronic signals usually have been generated and processed by integrated circuits.

1-2 WHERE ARE DIGITAL CIRCUITS USED?

Digital electronics is a fast-growing field, as witnessed by the pocket calculator, for one example. Figure 1-3 shows a specialized calculator produced by Texas Instruments, Inc. Now for just a few dollars you can purchase a simple pocket calculator. Only 15 to 20 years ago the same calculator might have cost thousands of dollars. Scientists, engineers, and technicians have made great advances in making integrated circuits. Due to these advances, the field of digital electronics has mushroomed.

Digital circuits have always been used in computers. Figure 1-4 is one version of a

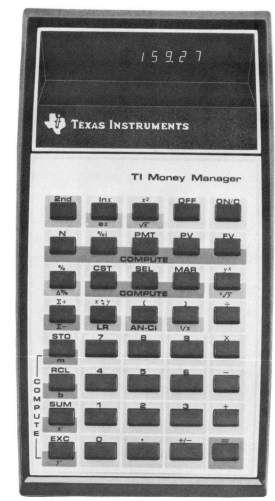

Fig. 1-3 **Personal calculator.** (*Courtesy of Texas Instruments, Inc.*)

build-it-yourself digital computer, a personal computer for home use. This unit is based upon an integrated circuit called a *microprocessor*. Units such as this one have started the personal computer revolution. Small computers that used to cost millions now cost only hundreds of dollars. Digital circuits are used in both large and small computers.

Fig. 1-4 **Personal digital computer.** (*Courtesy of Heath Company.*)

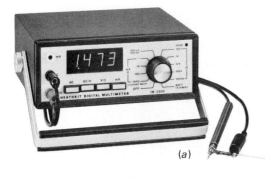

(a)

(b)

Fig. 1-5 Digital test equipment. (a) **Digital multimeter.** (b) **Frequency counter.** (*Courtesy of Heath Company.*)

(a)

The technician's bench has a new look. Digital *multimeters* silently read out resistance, voltage, and current values. Figure 1-5(a) illustrates one such digital multimeter. Digital multimeters are easy to use, extremely accurate, and modestly priced.

There is also a *frequency counter* on the technician's bench. This fantastic digital instrument accurately senses and reads out the frequency of alternating current (ac) signals into millions of cycles per second. Figure 1-5(b) pictures one such frequency counter. Both frequency counters and digital multimeters extensively use digital circuits. The oscilloscope also is beginning to make use of some digital circuits.

Digital electronic circuits are in your automobile. There may be a microprocessor to precisely adjust the spark advance on your car's engine. A digital tachometer stares back at you from the dashboard. Figure 1-6(a) shows a digital tachometer that you could construct. A digital clock and digital speedometer may be imbedded in the instrument panel of your automobile. Reddish digits glow from below the dashboard. Your mobile radio also uses some digital circuits. Figure 1-6(b) shows a CB radio with a digital readout. More and more digital circuits are appearing in radios and televisions.

Digital multimeters

Frequency counter

Digital tachometer

Digital clock

Digital speedometer

(b)

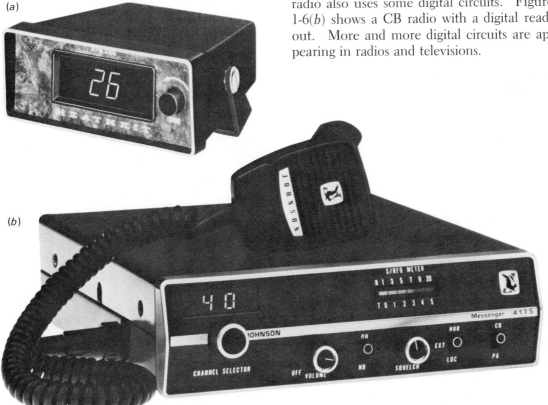

Fig. 1-6 (a) **Digital tachometer.** (*Courtesy of Heath Company.*)
(b) **CB Radio with digital readout.** (*Courtesy of E. F. Johnson Company.*)

3

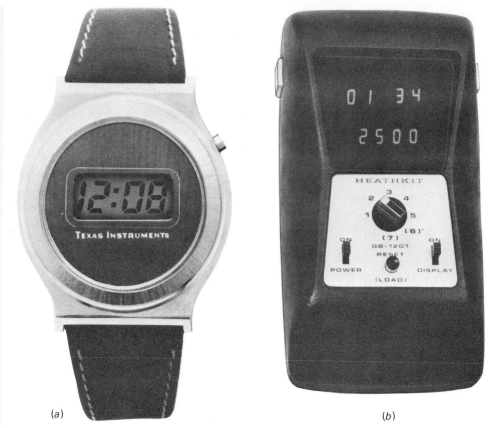

(a)

(b)

Fig. 1-7 (a) **Digital wristwatch.** (*Courtesy of Texas Instruments, Inc.*) (b) **Digital stopwatch.** (*Courtesy of Heath Company.*)

(a)

(b)

Fig. 1-8 (a) **Digital thermometer.** (b) **Digital AM/FM stereo receiver.** (*Courtesy of Heath Company.*)

Digital circuits are used in numerous items sold at your local shopping center. A digital wristwatch flashes the time in the jewelry section of a store. Figure 1-7(a) pictures one such watch. Digital watches, which are noted for their extreme accuracy, are a triumph of modern microelectronics technology. Another digital timepiece is shown in Fig. 1-7(b). The digital stopwatch has all but replaced the mechanical type. Both digital timepieces are based upon digital electronic circuits.

In your home there are also digital circuits. The temperature is flashed to you on a digital thermometer such as the one shown in Fig. 1-8(a). Some units also show the wind speed and direction. In the past, stereo and radio systems contained only analog circuits, but recently digital circuits have been used in these units. Figure 1-8(b) shows a digital AM/FM stereo receiver manufactured by Heath Company. The receiver still contains mostly analog circuits, but digital circuits are beginning to appear in these systems. Electronic games played on your television set are based upon digital electronic circuits.

Formerly digital circuits were used just in computers. Now these circuits are being widely used in many other products because of their low cost. Since digital circuits are so much at work in so many items, you will need to know how they operate.

Digital
wristwatch

Digital
thermometer

Electronic games

Questions

1. Define the following:
   a. Analog signal
   b. Digital signal

2. Draw a square-wave digital signal. Label the bottom 0 V and the top +5 V. Label the HIGH and LOW on the waveform. Label the logical 1 and logical 0 on the waveform.

3. List two devices that contain digital circuits which do mathematical calculations.

4. List three test instruments that contain digital circuits and are used by electronic technicians.

5. List three devices that contain some digital circuits and are used in many new automobiles.

6. List two types of timepieces that contain digital circuits.

7. Why are digital electronic circuits being more widely used?

Numbers We Use in Digital Electronics

■ Most people understand us when we say we have nine pennies. The number 9 is part of the *decimal* number system we use every day. But digital electronics devices use a "strange" number system called *binary*. Men and women who work in electronics must know how to convert numbers from the everyday decimal system to the binary system. By the end of this chapter you will be able to convert common decimal numbers to binary numbers and binary numbers to common decimal numbers.

2-1 COUNTING IN DECIMAL AND BINARY

A number system is a code that uses symbols to refer to a number of items. The decimal number system uses the symbols 0, 1, 2, 3, 4, 5, 6, 7, 8, and 9. The decimal number system contains 10 symbols and is sometimes called the *base 10 system*. The binary number system uses only the two symbols 0 and 1 and is sometimes called the *base 2 system*.

Figure 2-1 compares a number of coins with the symbols we use for counting. The decimal symbols that we commonly use for counting from 0 to 9 are shown in the left column;

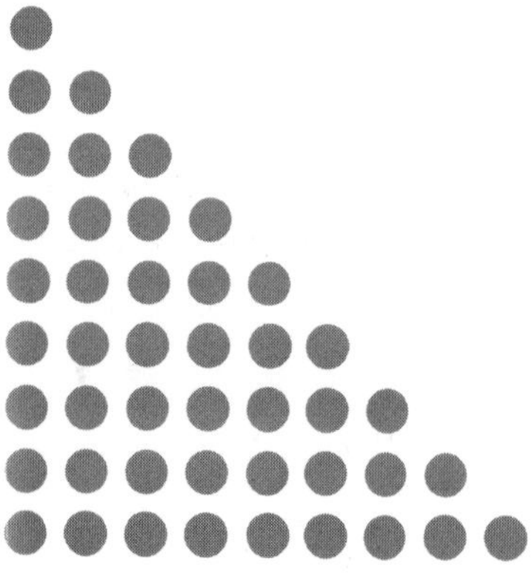

COINS	DECIMAL SYMBOL	BINARY SYMBOL
No coins	0	0
●	1	1
●●	2	10
●●●	3	11
●●●●	4	100
●●●●●	5	101
●●●●●●	6	110
●●●●●●●	7	111
●●●●●●●●	8	1000
●●●●●●●●●	9	1001

Fig. 2-1 Symbols for counting.

the right column has the symbols we use to count nine coins in the binary system. Notice that the 0 and 1 count in binary is the same as in decimal counting. To represent two coins, the binary number 10 (say "one zero") is used. To represent three coins, the binary number 11 (say "one one") is used. To represent nine coins, the binary number 1001 (say "one zero zero one") is used.

For your work in digital electronics you must at least memorize the binary symbols used to count up to 9.

2-2 PLACE VALUE

The clerk at the local restaurant totals your food bill and asks you for $2.43. We all know that this amount equals 243 cents. But instead of paying the clerk with 243 pennies, you probably would give the clerk the money shown in Fig. 2-2: two dollar bills, four dimes, and three pennies. This money example illustrates the very important idea of *place value*.

Consider the decimal number 648 in Fig. 2-3. The digit 6 represents 600 because of its placement three positions left of the decimal point. The digit 4 represents 40 because of its placement two positions left of the decimal point. The digit 8 represents eight units because of its placement one position left of the decimal point. The total number 648, then, represents six hundred and forty-eight units.

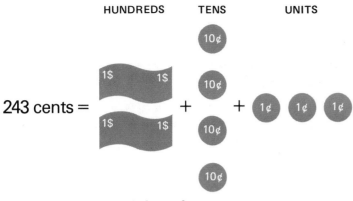

HUNDREDS TENS UNITS

243 cents =

Fig. 2-2 An example of place value.

This is an example of place value in the decimal number system.

	HUNDREDS		TENS		UNITS
648 =	600	+	40	+	8

Fig. 2-3 Place value in the decimal system.

The binary number system also uses the idea of place value. What does the binary number 1101 (say "one one zero one") mean? Figure 2-4 shows that the digit 1 nearest the binary point is the units or 1s position, so we add one item. The digit 0 in the 2s position tells us we have no 2s. The digit 1 in the 4s position tells us to add four items. The digit 1 in the 8s position tells us to add eight more items. When we count all the items, we find that the binary number 1101 represents 13 items.

How about the binary number 1100 (say "one one zero zero")? Using the system from Fig. 2-4, we find that we have the following:

8s	4s	2s	1s	place value
yes (1)	yes (1)	no (0)	no (0)	binary number
•• •• •• ••	•• ••			number of items

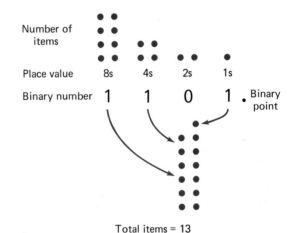

Fig. 2-4 Place value in the binary number system.

The binary number 1100, then, represents 12 items.

Figure 2-5 shows how much each place value is worth in the binary system. Notice that each place value is determined by multiplying the one to the right by 2. The base 2 term for binary comes from this idea.

2-3 BINARY TO DECIMAL CONVERSION

While working with digital equipment you will have to convert from the binary code to decimal numbers. If you are given the binary

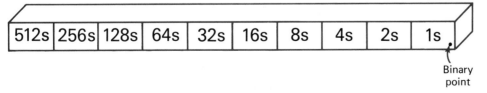

512s	256s	128s	64s	32s	16s	8s	4s	2s	1s

Binary point

Fig. 2-5 Values of the places left of the binary point.

Decimal to binary conversion

number 110011, what would it equal in decimals? First write down the binary number as:

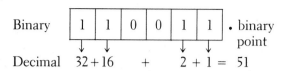

Binary | 1 | 1 | 0 | 0 | 1 | 1 | . binary point
Decimal 32 + 16 + 2 + 1 = 51

Start at the binary point and work to the left. For each binary 1, place the decimal value of that position (see Fig. 2-5) below the binary digit. Add the four decimal numbers to find the decimal equivalent. You will find that binary 110011 equals the decimal number 51.

Another practical problem is to convert the binary number 101010 to a decimal number. Again write down the binary number as:

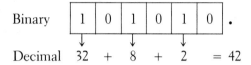

Binary | 1 | 0 | 1 | 0 | 1 | 0 | .
Decimal 32 + 8 + 2   = 42

Starting at the binary point, write the place value (see Fig. 2-5) of each digit below the squares in decimals. Add the three decimal numbers to get the decimal total. You will find that the binary number 101010 equals the decimal number 42.

Now try a long and difficult binary number: convert the binary number 1111101000 to a decimal number. Write down the binary number as:

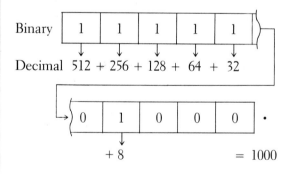

Binary | 1 | 1 | 1 | 1 | 1 |
Decimal 512 + 256 + 128 + 64 + 32

| 0 | 1 | 0 | 0 | 0 | .
+ 8   = 1000

From Fig. 2-5, convert each binary digit 1 into its correct decimal value. Add the decimal values to get the decimal total. The binary number 1111101000 equals the decimal number 1000.

2-4 DECIMAL TO BINARY CONVERSION

Many times while working with digital electronic equipment you must be able to convert a decimal number into a binary number. We shall teach you a method that will help you to make this conversion.

Suppose you want to convert the decimal number 13 to a binary number. One procedure you can use is:

Decimal number

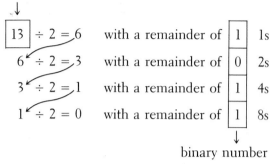

$13 \div 2 = 6$ with a remainder of | 1 | 1s
$6 \div 2 = 3$ with a remainder of | 0 | 2s
$3 \div 2 = 1$ with a remainder of | 1 | 4s
$1 \div 2 = 0$ with a remainder of | 1 | 8s

binary number

Notice that 13 is first divided by 2, giving a quotient of 6 with a remainder of 1. This remainder becomes the 1s place in the binary number. The 6 is then divided by 2, giving a quotient of 3 with a remainder of 0. This remainder becomes the 2s place in the binary number. The 3 is then divided by 2, giving a quotient of 1 with a remainder of 1. This remainder becomes the 4s place in the binary number. The 1 is then divided by 2, giving a quotient of 0 with a remainder of 1. This remainder becomes the 8s place in the binary number. The decimal 13 has been converted to the binary number 1101.

Practice this procedure by converting the decimal number 37 to a binary number. Follow the procedure you used before:

Decimal number

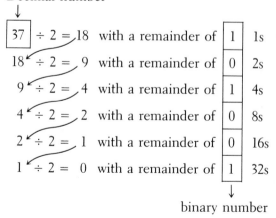

$37 \div 2 = 18$ with a remainder of | 1 | 1s
$18 \div 2 = 9$ with a remainder of | 0 | 2s
$9 \div 2 = 4$ with a remainder of | 1 | 4s
$4 \div 2 = 2$ with a remainder of | 0 | 8s
$2 \div 2 = 1$ with a remainder of | 0 | 16s
$1 \div 2 = 0$ with a remainder of | 1 | 32s

binary number

Notice that you stop dividing by 2 when the quotient becomes 0. According to this procedure, the decimal number 37 is equal to the binary number 100101.

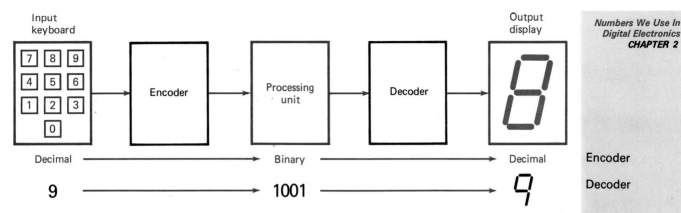

Fig. 2-6   A system using encoders and decoders.

Self Test

Check your understanding by answering questions 1 and 2.

1. Convert the binary numbers in *a* to *f* to decimal numbers:
   a. 101   *d.* 10010
   b. 1001   *e.* 111
   c. 1111   *f.* 1000001010

2. Convert the decimal numbers in *a* to *f* to binary numbers:
   a. 8   *d.* 34
   b. 11   *e.* 100
   c. 17   *f.* 133

2-5 ELECTRONIC TRANSLATORS

If you were to try to communicate with a French person who did not know the English language, you would need someone to *translate* the English into French and then the French into English. A similar problem exists in digital electronics. Almost all digital circuits (calculators, computers) understand only binary numbers. But most people understand only decimal numbers. Thus we must have electronic devices that can translate from decimal to binary numbers and from binary to decimal numbers.

Figure 2-6 diagrams a typical system that might be used to translate from decimal to binary numbers and back to decimals. The device that translates from the keyboard decimal numbers to binary is called an *encoder*; the device labeled *decoder* translates from binary to decimal numbers.

Near the bottom of Fig. 2-6 is a typical conversion. If you press the decimal number 9 on the keyboard, the encoder will convert the 9 into the binary number 1001. The decoder will convert the binary 1001 into the decimal number 9 on the output display.

Encoders and decoders are very common electronic circuits in all digital devices. A pocket calculator, for instance, must have encoders and decoders to electronically translate from decimal to binary numbers and back to decimals. Figure 2-6 is a very basic block diagram of a pocket calculator when you press the number 9 on the keyboard.

You can buy encoders and decoders that translate from any of the commonly used codes in digital electronics. Most of the encoders and decoders you will use will be packaged as single integrated circuits.

Summary

1. The decimal number system contains 10 symbols: 0, 1, 2, 3, 4, 5, 6, 7, 8, and 9.
2. The binary number system contains two symbols: 0 and 1.
3. The place values left of the binary point in binary are 64, 32, 16, 8, 4, 2, and 1.
4. All men and women in the field of digital electronics must be able to convert binary to decimal numbers and decimal to binary numbers.
5. Encoders are electronic circuits that convert decimal numbers to binary numbers.
6. Decoders are electronic circuits that convert binary numbers to decimal numbers.

Questions

1. How would you say the decimal number 1001?

2. How would you say the binary number 1001?

3. Convert the binary numbers in *a* to *j* into decimal numbers:
   a. 1   *f.* 10000
   b. 100   *g.* 10101
   c. 101   *h.* 11111
   d. 1011   *i.* 11001100
   e. 1000   *j.* 11111111

4. Convert the decimal numbers in *a* to *j* into binary numbers:
   a. 0   *f.* 64
   b. 1   *g.* 69
   c. 18   *h.* 128
   d. 25   *i.* 145
   e. 32   *j.* 1001

5. Encode the decimal numbers in *a* to *f* to binary numbers:
   a. 9   *d.* 13
   b. 3   *e.* 10
   c. 15   *f.* 2

6. Decode the binary numbers in *a* to *f* to decimal numbers:
   a. 0010   *d.* 0111
   b. 1011   *e.* 0110
   c. 1110   *f.* 0000

7. What is the job (function) of an encoder?

8. What is the job (function) of a decoder?

9. Write the numbers from 0 to 15 in binary.

Answers to Self Test

1. *a.* 5   2. *a.* 1000
   b. 9   *b.* 1011
   c. 15   *c.* 10001
   d. 18   *d.* 100010
   e. 7   *e.* 1100100
   f. 524   *f.* 10000101

Binary Logic Gates

- Computers, calculators, and other digital devices are sometimes looked upon by the general public as being magical. Actually, digital electronic devices are extremely *logical* in their operation. The basic building block of any digital circuit is a *logic gate.* The logic gates you will use operate with binary numbers, hence the term *binary logic gates.*

  Persons working in digital electronics understand and use binary logic gates every day. By the end of this chapter you will have memorized the basic logic gate symbols and characteristics. Remember that logic gates are the building blocks for even the most complex computers.

  Logic gates can be constructed by using simple switches, relays, vacuum tubes, transistors and diodes, or integrated circuits. Because of their availability, wide use, and low cost, we shall use *integrated circuits (ICs)* in our digital circuits.

3-1 THE AND GATE

The AND gate is sometimes called the "all or nothing gate." Figure 3-1 shows the basic idea of the AND gate using simple switches.

What must be done in Fig. 3-1 to get the output lamp (L_1) to light? You must close *both* switches A and B to get the lamp to light. You could say that switch A *and* switch B must be closed to get the output to light.

The AND gates you will operate most often are constructed of diodes and transistors and packaged inside an integrated circuit (IC). To show the AND gate we use the *logic symbol* in Fig. 3-2. This standard AND gate sym-

Fig. 3-2   AND gate logic symbol.

bol is used whether we are using relays, switches, pneumatic circuits, discrete diodes and transistors, or ICs. This is the symbol you will memorize and use from now on for AND gates.

The term *logic* is usually used to refer to a decision-making process. A logic gate, then, is a circuit that can decide to say yes or no at the output based upon the inputs. We already determined that the AND gate circuit in Fig. 3-1 says Yes (light on) at the output only when we have a Yes (switches closed) at *both* inputs.

Now let us consider an actual circuit similar to one you will set up in the laboratory. The AND gate in Fig. 3-3 is connected to input switches A and B. The output indicator is a light-emitting diode (LED). If a LOW voltage (GND) appears at inputs A and B, then the output LED is *not lit*. This situation is illustrated in line 1, Fig. 3-4. Notice also in line 1 that the inputs and output are given as a *binary digit*. Line 1 indicates that if the

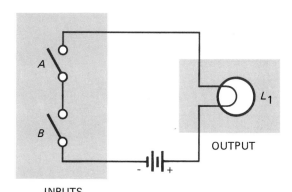

INPUTS

Fig. 3-1   AND circuit using switches.

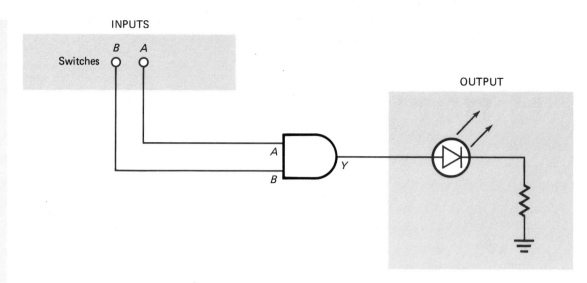

INPUTS

OUTPUT

Fig. 3-3   Practical AND gate circuit.

inputs are binary 0 and 0, then the output will be a binary 0. Carefully look over the four combinations of switches A and B in Fig. 3-4. Notice that only binary 1s at both inputs A and B will produce a binary 1 at the output (see line 4).

It is a +5 V compared to ground (GND) appearing at A, B, or Y that is called a binary 1 or a HIGH voltage. A binary 0 or LOW voltage is defined as a ground voltage (near 0 V compared to GND) appearing at A, B, or Y. We are using *positive logic* because it takes a *positive* 5 V to produce what we call a binary 1. You will use positive logic in most of your work in digital electronics.

The table in Fig. 3-4 is called a *truth table*. The truth table for the AND gate gives all the possible input combinations of A and B and the resulting outputs. Thus the truth table defines very exactly the operation of the AND gate. You will memorize the truth table for the AND gate.

So far you have memorized the logic sym-

bol and the truth table for the AND gate. Now you will learn a shorthand method of writing the statement "input A is ANDed with input B to get output Y." The short method for this statement is called its *Boolean expression* (Boolean from Boolean algebra—the algebra of logic). The Boolean expression is a universal language used by engineers and technicians in digital electronics. Figure 3-5 shows the ways to express that input A is ANDed with input B to produce output Y. The top expression in Fig. 3-5 is how you would tell someone in the English language that you are ANDing inputs A and B to get output Y. Next in Fig. 3-5 you see the Boolean expression for ANDing inputs A and B. Note that a multiplication dot (·) is used to symbolize the AND function in Boolean expressions. Figure 3-5, then, illustrates the four commonly used ways to express the ANDing of inputs A and B. All these methods are widely used and must be learned by personnel in digital electronics.

	INPUTS				OUTPUT	
	B		A		Y	
	Switch voltage	Binary	Switch voltage	Binary	Light	Binary
Line 1	LOW	0	LOW	0	No	0
Line 2	LOW	0	HIGH	1	No	0
Line 3	HIGH	1	LOW	0	No	0
Line 4	HIGH	1	HIGH	1	Yes	1

Fig. 3-4   AND gate truth table.

In the English language	Input *A* is ANDed with input *B* to get output *Y*.
As a Boolean expression	$A \cdot B = Y$ AND symbol
As a logic symbol	 *A* *B* → *Y*
As a truth table	*B* *A* *Y* 0 0 0 0 1 0 1 0 0 1 1 1

Fig. 3-5   Four ways to express the logical ANDing of *A* and *B*.

INPUTS				OUTPUT	
B		A		Y	
Switch	Binary	Switch	Binary	Light	Binary
Open	0	Open	0	No	0
Open	0	Closed	1	Yes	1
Closed	1	Open	0	Yes	1
Closed	1	Closed	1	Yes	1

Fig. 3-7   OR gate truth table.

3-2 THE OR GATE

The OR gate is sometimes called the "any or all gate." Figure 3-6 illustrates the basic idea of the OR gate using simple switches. Looking at the circuit in Fig. 3-6, you can see that the output lamp will light when *either or both* of the input switches are closed but not when both are open. A truth table for the OR circuit is shown in Fig. 3-7. The truth table lists the switch and light conditions for the OR gate circuit in Fig. 3-6.

The logic symbol for the OR gate is diagramed in Fig. 3-8. Notice in the logic diagram that inputs *A* and *B* are being ORed to produce an output *Y*. The engineer's Boolean expression for the OR function is also illustrated in Fig. 3-8. Note that the plus (+) sign is the Boolean symbol for OR.

You should now know the logic symbol, Boolean expression, and truth table for the OR gate.

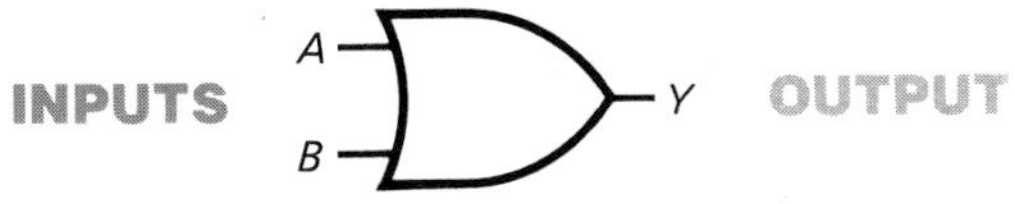

INPUTS   *A* *B* → *Y*   OUTPUT

$$A + B = Y$$

OR symbol

Fig. 3-8   OR gate logic symbol and a Boolean expression.

3-3 THE INVERTER

All the gates so far have had at least two inputs and one output. The NOT circuit, however, has only one input and one output. The NOT circuit is often called an *inverter*. The job of the NOT circuit (inverter) is to give an output that is not the same as the input. The logic symbol for the inverter (NOT circuit) is shown in Fig. 3-9.

If we were to put in a logical 1 at input *A* in Fig. 3-9, we would get out the opposite, or a logical 0, at output *Y*. We say that the inverter *complements* or *inverts* the input. Figure 3-9 also shows how we would write a Boolean expression for the NOT or INVERT function. Notice the use of the bar (—) symbol

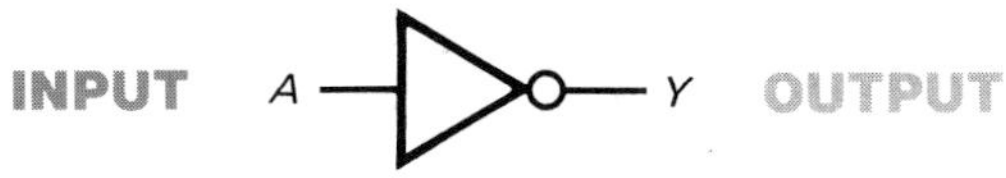

INPUT   *A* → *Y*   OUTPUT

$$A = \overline{A}$$

NOT symbol

Fig. 3-9   A logic symbol and Boolean expression for an inverter.

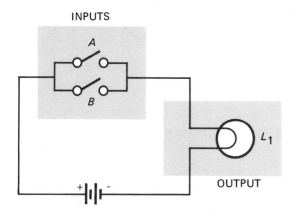

INPUTS

A

B

L_1

OUTPUT

Fig. 3-6   OR circuit using switches.

INPUT		OUTPUT	
A		Y	
Voltages	Binary	Voltages	Binary
LOW	0	HIGH	1
HIGH	1	LOW	0

Fig. 3-10   Truth table for an inverter.

Complemented

Negated

NAND gate

above the output to show that A has been inverted or complemented. We say that the Boolean term $\bar{A}$ would be "not A."

The truth table for the inverter is shown in Fig. 3-10. If the voltage at the input of the inverter is low, then the output voltage is high. However, if the input voltage is high, then the output will be low. As you learned, the output is always opposite the input. The truth table also gives the characteristics of the inverter in terms of binary 0s and 1s.

You learned that when a signal passes through an inverter, it can be said that the input is inverted or complemented. We can also say it is *negated*. The terms negated, complemented, and inverted, then, mean the same thing.

The logic diagram in Fig. 3-11 shows an arrangment where input A is run through two inverters. Input A is first inverted to produce a "not A" ($\bar{A}$) and then inverted a second time for a "double not A" ($\bar{\bar{A}}$). In terms of binary digits, we find that when the input 1 is inverted twice, we end up with the original digit. Therefore we find that $\bar{\bar{A}}$ equals A. A Boolean term with two bars over it is equal to the term under the two bars, as shown at the bottom of Fig. 3-11.

You now know the logic symbol, Boolean expression, and truth table for the inverter. You will use these items every day as you work in the digital electronics field.

Self Test

Check your understanding by answering questions 1 to 5.

1. Draw the logic symbol for the following:
   a. OR gate
   b. AND gate
   c. inverter

2. Write the Boolean expression for *a* to *c* (in *a* and *b* use letters A and B as inputs and Y for outputs):
   a. OR function
   b. AND function
   c. NOT function (use A for input and output)

3. Make a truth table for *a* to *c* (use binary 0s and 1s):
   a. OR gate
   b. AND gate
   c. inverter

4. List two other words that are used to mean *inverted*.

5. The Boolean term $\bar{\bar{B}} = $ ___?___.

3-4 THE NAND GATE

The AND, OR, and NOT gates are the three basic circuits that make up all digital circuits. The NAND gate is a NOT AND or an inverted AND function. The standard logic symbol for the NAND gate is diagramed in Fig. 3-12(a). The little bubble (small circle) on the right end of the symbol means to invert the AND.

Figure 3-12(b) shows a separate AND gate and inverter being used to produce the NAND logic function. Also notice that the

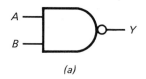

(a)

INPUT   **OUTPUT**

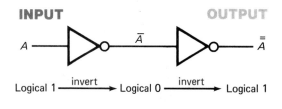

Logical 1 →invert→ Logical 0 →invert→ Logical 1

Therefore $\bar{\bar{A}} = A$

Fig. 3-11   Double inverting.

INPUTS   **OUTPUT**

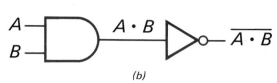

(b)

Fig. 3-12   (a) NAND gate logic symbol. (b) Boolean expression for the output of a NAND gate.

INPUTS		OUTPUT	
B	A	AND	NAND
0	0	0	1
0	1	0	1
1	0	0	1
1	1	1	0

Fig. 3-13 Truth tables for AND and NAND gates.

INPUTS		OUTPUT	
B	A	OR	NOR
0	0	0	1
0	1	1	0
1	0	1	0
1	1	1	0

Fig. 3-15 Truth table for OR and NOR gates.

NOR gate

Exclusive OR gate

Boolean expression for the AND gate $(A \cdot B)$ and the NAND $(\overline{A \cdot B})$ is shown on the logic diagram in Fig. 3-12(*b*).

The truth table for the NAND gate is shown at the right in Fig. 3-13. Notice that the truth table for the NAND gate is developed by *inverting* the outputs of the AND gate. The AND gate outputs are also given in the table for reference.

NAND gates are commonly employed in industrial practice and extensively used in all digital equipment. Do you know the logic symbol, Boolean expression, and truth table for the NAND gate? You must commit these to memory.

3-5 THE NOR GATE

The NOR gate is actually a NOT OR gate. In other words, the output of an OR gate is inverted to form a NOR gate. The logic symbol for the NOR gate is diagramed in Fig. 3-14(*a*). Note that the NOR symbol is an OR symbol with a small *invert bubble* (small circle) on the right side. The NOR function is being performed by an OR gate and an inverter in Fig. 3-14(*b*). The Boolean expres-

sion for the OR function $(A + B)$ is shown. The final NOR function is given the Boolean expression $\overline{A + B}$.

The truth table for the NOR gate is shown at the right in Fig. 3-15. Notice that the NOR gate truth table is just the complement of the output of the OR gate. The output of the OR gate is also included in the truth table in Fig. 3-15 for reference.

You now should memorize the symbol, Boolean expression, and truth table for the NOR gate. You will encounter these items often in your work in digital electronics.

3-6 THE EXCLUSIVE OR GATE

The exclusive OR gate is sometimes referred to as the "any but not all gate." The exclusive OR gate is often shortened to the XOR gate. The logic symbol for the XOR gate is diagramed in Fig. 3-16(*a*); the Boolean expression for the XOR function is illustrated in Fig. 3-16(*b*). The symbol $\oplus$ means the terms are XORed together.

A truth table for the XOR gate is shown at the right in Fig. 3-17. Notice that if any but

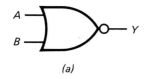

(a)

INPUTS   **OUTPUT**

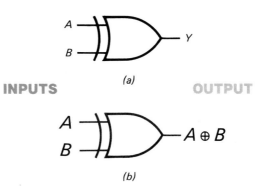

(b)

Fig. 3-14 (*a*) NOR gate logic symbol. (*b*) A Boolean expression for the output of a NOR gate.

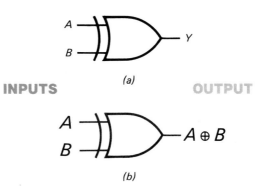

(a)

INPUTS   **OUTPUT**

(b)

Fig. 3-16 (*a*) XOR gate logic symbol. (*b*) A Boolean expression for the output of an XOR gate.

15

INPUTS		OUTPUT	
B	A	OR	XOR
0	0	0	0
0	1	1	1
1	0	1	1
1	1	1	0

Fig. 3-17   Truth table for OR and XOR gates.

INPUTS		OUTPUT	
B	A	XOR	XNOR
0	0	0	1
0	1	1	0
1	0	1	0
1	1	0	1

Fig. 3-19   Truth table for XOR and XNOR gates.

Exclusive NOR gate

Universal gate

not all of the inputs are 1, then the output will be a binary or logical 1. The OR gate truth table is also given in Fig. 3-17 so you may compare the OR gate truth table with the XOR gate truth table.

3-7 THE EXCLUSIVE NOR GATE

The exclusive NOR gate is often shortened to the XNOR gate. The logic symbol for the XNOR gate is shown in Fig. 3-18(a). Notice that it is the XOR symbol with the added invert bubble on the output side. Figure 3-18(b) illustrates one of the Boolean expressions used for the XNOR function. Observe that the Boolean expression for the XNOR gate is $\overline{A \oplus B}$. The bar over the $A \oplus B$ expression tells us we have an inverted XOR gate. Examine the truth table in Fig. 3-19. Notice that the output of the XNOR gate is the complement of the XOR truth table. The XOR gate output is also shown in the table in Fig. 3-19 for your convenience.

You now will have mastered the logic symbol, truth table, and Boolean expression for the XNOR gate.

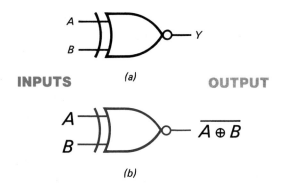

Fig. 3-18   (a) XNOR gate logic symbol. (b) A Boolean expression for the output of an XNOR gate.

Self Test

Check your understanding by answering questions 6 to 8.

6. Draw a logic symbol for *a* to *d*:
   a. NAND gate   *c*. XOR gate
   b. NOR gate   *d*. XNOR gate

7. Write the Boolean expression for *a* to *d* (use A and B as inputs and Y as an output):
   a. NAND function   *c*. XOR function
   b. NOR function   *d*. XNOR function

8. Make a truth table for *a* to *d* (use logical 0s and 1s):
   a. NAND gate   *c*. XOR gate
   b. NOR gate   *d*. XNOR gate

3-8 THE NAND GATE AS A UNIVERSAL GATE

So far in this chapter you have learned the basic building blocks used in all digital circuits. You also have learned about the seven types of gating circuits and now know the characteristics of the AND, OR, NAND, NOR, XOR, and XNOR gates and the inverter. From a manufacturer you can buy ICs that perform any of these seven basic functions.

In looking through manufacturers' literature you will find that NAND gates seem to be more widely available than many other types of gates. Because of the NAND gate's wide use, we shall show how it can be used to make other types of gates. We shall be using the NAND gate as a sort of "universal gate."

The table in Fig. 3-20 shows how you would wire NAND gates to create any of the other basic logic functions. The logic function to be performed is listed in the left column of the table; the symbol for that function is listed in the center column. In the right column of Fig. 3-20 is a symbol diagram of how NAND

LOGIC FUNCTION	SYMBOL	CIRCUIT USING NAND GATES ONLY
Inverter	A —▷∘— $\bar{A}$	
AND	$A \cdot B$	$A \cdot B$
OR	$A + B$	$A + B$
NOR	$\overline{A + B}$	$\overline{A + B}$
XOR	$A \oplus B$	$A \oplus B$
XNOR	$\overline{A \oplus B}$	$\overline{A \oplus B}$

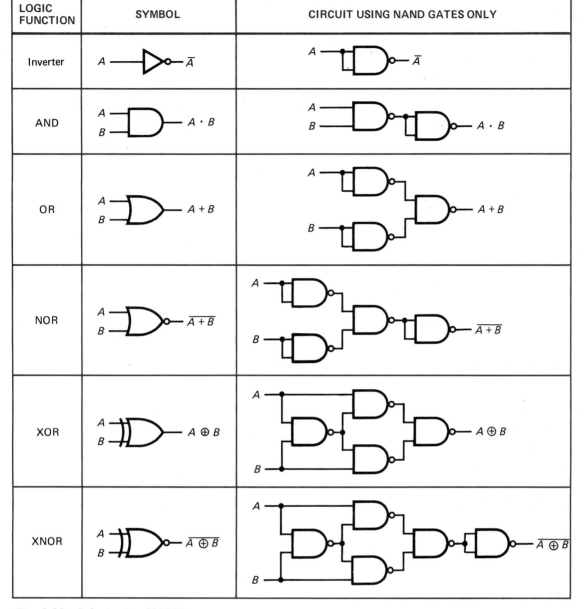

Fig. 3-20   Substituting **NAND** gates.

3-input AND gate

gates would be wired to perform the logic function. The chart in Fig. 3-20 need *not* be memorized, but it may be referred to as needed in your future work in digital electronics.

3-9 GATES WITH MORE INPUTS

Until now we have used gates that have had only two or less inputs. But often we shall need gates with more than just two inputs. Figure 3-21(*a*) shows a 3-input AND gate. The Boolean expression for the 3-input AND

gate is $A \cdot B \cdot C = Y$, as illustrated in Fig. 3-21(*b*). All the possible combinations of inputs A, B, and C are given in the truth table in Fig. 3-21(*c*); the outputs for the 3-input AND gate are tabulated in the right column of the truth table. Notice that with three inputs the possible combinations in the truth table have increased to eight.

How could you produce a 3-input AND gate as illustrated in Fig. 3-21 if you have only 2-input AND gates available? The solution is given in Fig. 3-22(*a*). Note the wiring of the 2-input AND gates on the right side of the diagram to form a 3-input AND gate. Figure

17

4-input OR gate

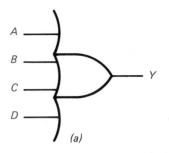

$$A \cdot B \cdot C = Y$$

(b)

INPUTS			OUTPUTS
C	B	A	Y
0	0	0	0
0	0	1	0
0	1	0	0
0	1	1	0
1	0	0	0
1	0	1	0
1	1	0	0
1	1	1	1

(c)

Fig. 3-21 3-input AND gate. (a) Logic symbol.
(b) Boolean expression. (c) Truth table.

3-22(b) illustrates how a 4-input AND gate
could be wired by using available 2-input
AND gates.

A 4-input OR gate is illustrated in Fig.
3-23(a). The Boolean expression for the
4-input OR gate is $A + B + C + D = Y$.
This Boolean expression is written in Fig.
3-23(b). Read the Boolean expression

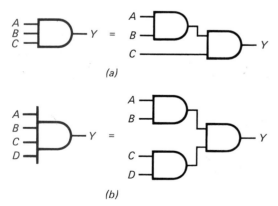

(a)

(b)

Fig. 3-22 Expanding the number of inputs.
Note the method used in (b) to accommodate the
extra inputs beyond the width of the symbol.

$$A + B + C + D = Y$$

(b)

INPUTS				OUTPUT
D	C	B	A	Y
0	0	0	0	0
0	0	0	1	1
0	0	1	0	1
0	0	1	1	1
0	1	0	0	1
0	1	0	1	1
0	1	1	0	1
0	1	1	1	1
1	0	0	0	1
1	0	0	1	1
1	0	1	0	1
1	0	1	1	1
1	1	0	0	1
1	1	0	1	1
1	1	1	0	1
1	1	1	1	1

(c)

Fig. 3-23 4-input OR gate. (a) Logic symbol
showing the method used to accommodate extra
inputs beyond the width of the symbol. (b) Boolean expression. (c) Truth table.

$A + B + C + D = Y$ as "input A or input B
or input C or input D will equal output Y."
Remember that the + symbol means the logic
function OR in Boolean expressions. The
truth table for the 4-input OR gate is shown
in Fig. 3-23(c). Notice that because of the
four inputs there are 16 possible combinations
of A, B, C, and D. To wire the 4-input OR
gate, you could buy the correct gate from a
manufacturer of digital logic circuits, or you
could wire the 4-input OR gate. Figure 3-24(a)
diagrams how you could wire a 4-input OR

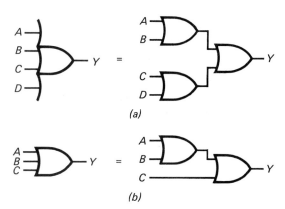

Fig. 3-24   Expanding the number of inputs.

gate using 2-input OR gates.   Figure 3-24(*b*) shows how to convert 2-input OR gates into a 3-input OR gate.   Notice that the *pattern* of connecting both OR and AND gates to expand the number of inputs is the same (compare Figs. 3-22 and 3-24).

Expanding the number of inputs of a NAND gate is somewhat more difficult than expanding AND and OR gates.   Figure 3-25 shows how a 4-input NAND gate can be wired using two 2-input NAND gates and one 2-input OR gate.

You frequently will run into gates that have from two to as many as eight and more inputs.   The basics covered in this section are a handy reference when you need to expand the number of inputs to a gate.

over the chart: notice that in the top section only the outputs are inverted.   Inverting the outputs leads to rather predictable results, shown on the right side of the chart.

The center section of the chart shows only the gate inputs being inverted.   For instance, if you invert both inputs of an OR gate, the gate will become a NAND gate.   This fact is emphasized in Fig. 3-26(*a*).   Notice that the invert bubbles have been added to the OR gate in Fig. 3-26(*a*), which converts the OR gate to a NAND gate.   Also in the center section of the chart the inputs of the AND gate are being inverted.   This is redrawn in Fig. 3-26(*b*).   Notice that the invert bubbles at the input of the AND gate convert it into a NOR gate.   The new symbols at the left (with the input invert bubbles) in Fig. 3-26 are used in some logic diagrams in place of the more standard NAND and NOR logic symbols at the right.   Be aware of these new symbols because you will run into them in your future work in digital electronics.

The bottom section of the chart in Fig. 3-27 shows both the inputs and outputs being inverted.   Notice that by using inverters at the inputs and outputs you can convert back and forth from AND to OR and from NAND to NOR.

With the 12 conversions shown in the chart in Fig. 3-27 you can convert any basic gate (AND, OR, NAND, and NOR) to any other gate with just the use of inverters.   You will *not* need to memorize the chart in Fig. 3-27, but remember it for reference.

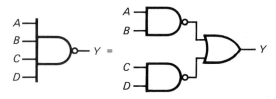

Fig. 3-25   Expanding the number of inputs.

3-10 CONVERTING GATES USING INVERTERS

Frequently it is convenient to convert a basic gate such as an AND, OR, NAND, or NOR to another logic function.   This can be done easily with the use of inverters.   The chart in Fig. 3-27 is a handy guide for converting any given gate into any other logic function.   Look

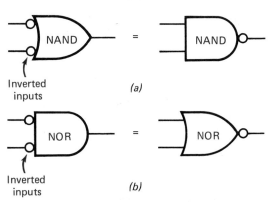

Fig. 3-26   (*a*) **NAND symbols.**   (*b*) **NOR symbols.**

19

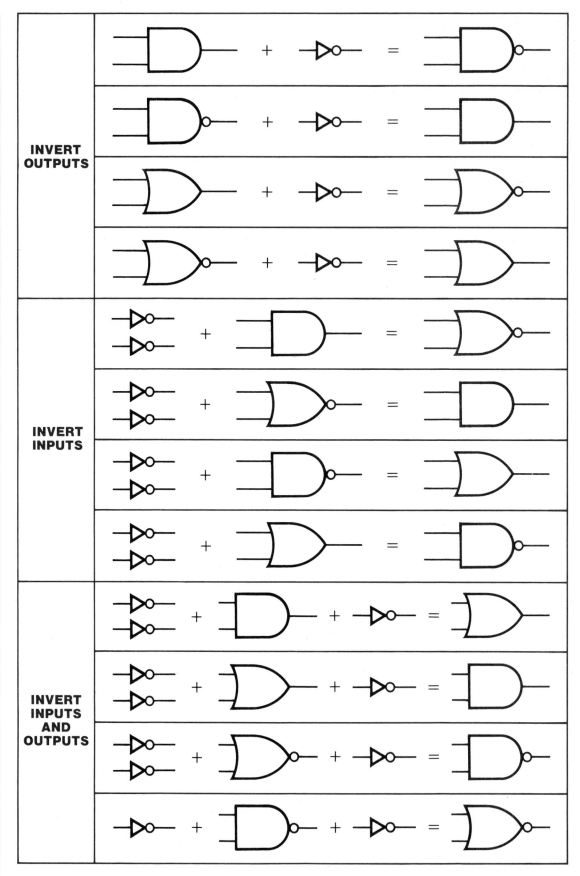

Fig. 3-27   Gate conversions using inverters. The + symbol here indicates adding the functions.

Summary

1. Binary logic gates are the basic building blocks for all digital circuits.

2. Figure 3-28 shows a summary of the seven basic logic gates. This information should be memorized.

3. NAND gates are very widely employed and can be used to make other logic gates.

4. Logic gates are often needed with 2 to 10 inputs. Several gates may be connected in the proper manner to get more inputs.

5. AND, OR, NAND, and NOR gates can be converted back and forth by using inverters.

LOGIC FUNCTION	LOGIC SYMBOL	BOOLEAN EXPRESSION	TRUTH TABLE		
			INPUTS		OUTPUT
			B	*A*	*Y*
AND	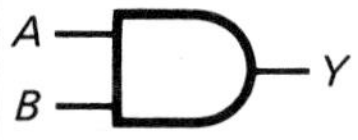	$A \cdot B = Y$	0	0	0
			0	1	0
			1	0	0
			1	1	1
OR	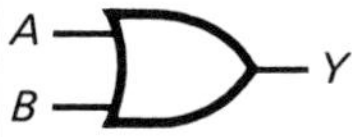	$A + B = Y$	0	0	0
			0	1	1
			1	0	1
			1	1	1
Inverter		$A = \overline{A}$		0	1
				1	0
NAND		$\overline{A \cdot B} = Y$	0	0	1
			0	1	1
			1	0	1
			1	1	0
NOR		$\overline{A + B} = Y$	0	0	1
			0	1	0
			1	0	0
			1	1	0
XOR		$A \oplus B = Y$	0	0	0
			0	1	1
			1	0	1
			1	1	0
XNOR	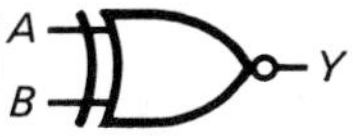	$\overline{A \oplus B} = Y$	0	0	1
			0	1	0
			1	0	0
			1	1	1

Fig. 3-28   Summary of basic logic gates.

Questions

1. Draw the logic symbols for *a* to *i* (label inputs A, B, C, D and outputs Y):
   a. 2-input AND gate
   b. 3-input OR gate
   c. Inverter
   d. 2-input XOR gate
   e. 4-input NAND gate
   f. 2-input NOR gate
   g. 2-input XNOR gate
   h. 2-input NAND gate (special symbol)
   i. 2-input NOR gate (special symbol)

2. Write the Boolean expression for each gate in question 1.

3. Write the truth table for each gate in question 1.

4. Look at the chart in Fig. 3-28. Which logic gate always responds with an output of logical 1 *just* when the two inputs are *unlike* (0 and 1 or 1 and 0)?

5. Which logic gate might be called the all or nothing gate?

6. Which logic gate might be called the any or all gate?

7. Which logic circuit *complements* the input?

8. Which logic gate might be called the any but not all gate?

9. Given an AND gate and inverters, draw how you would produce a NOR function.

10. Given a NAND gate and inverters, draw how you would produce an OR function.

11. Given a NAND gate and inverters, draw how you would produce an AND function.

12. Given four 2-input AND gates, draw how you would produce a 5-input AND gate.

13. Given several 2-input NAND and OR gates, draw how you would produce a 4-input NAND gate.

14. Switches arranged in series (see Fig. 3-1) act like what type of logic gate?

15. Switches arranged in parallel (see Fig. 3-6) act like what type of logic gate?

Answers to Self Tests

1. a.

   b.

   c. ⊳○

2. a. $A + B = Y$
   b. $A \cdot B = Y$
   c. $A = \overline{A}$

3. a.

B	A	Y
0	0	0
0	1	1
1	0	1
1	1	1

   b.

B	A	Y
0	0	0
0	1	0
1	0	0
1	1	1

   c.

A	Y
0	1
1	0

4. Complemented, negated

5. B

6. a.

   b.

   c.

   d.

7. *a.* $\overline{A \cdot B} = Y$
   b. $\overline{A + B} = Y$
   c. $A \oplus B = Y$
   d. $\overline{A \oplus B} = Y$

8. *a.*

B	A	Y
0	0	1
0	1	1
1	0	1
1	1	0

b.

B	A	Y
0	0	1
0	1	0
1	0	0
1	1	0

c.

B	A	Y
0	0	0
0	1	1
1	0	1
1	1	0

d.

B	A	Y
0	0	1
0	1	0
1	0	0
1	1	1

Using Binary Logic Gates

■ In Chap. 3 you memorized the symbol, truth table, and Boolean expression for each of the binary logic gates. These gates are the basic building blocks for *all* digital systems.

In this chapter you will see how your knowledge of gate symbols, truth tables, and Boolean expressions can be used to solve real-world problems in electronics. You will be connecting together gates to form what engineers refer to as *combinational logic circuits*, and you will be combining gates (ANDs, ORs) and inverters and so on to solve logic problems.

There are three "tools of the trade" in solving logic problems: gate *symbols*, *truth tables*, and *Boolean expressions*. Do you have these tools of the trade? Do you know your gate symbols, truth tables, and Boolean expressions? If you need review on logic gates, please refer to the summary, Chap. 3; Fig. 3-28 especially will be helpful.

Your understanding of using logic gates is important because to be sucessful as a technician, troubleshooter, engineer, or hobbyist in digital electronics you must master combining gates. It is highly suggested that you try out your *combinational logic circuits* in the shop or laboratory. Logic gates are packaged in inexpensive, easy-to-use ICs.

4-1 CONSTRUCTING CIRCUITS FROM BOOLEAN EXPRESSIONS

We use Boolean expressions to guide us in building logic circuits. Suppose you are given the Boolean expression $A + B + C = Y$ (read as "A or B or C equals output Y") and

told to build a logic circuit that would perform this logic. Looking at the expression, notice that each input must be ORed to get output Y. Figure 4-1 illustrates the *gate* needed to do the job.

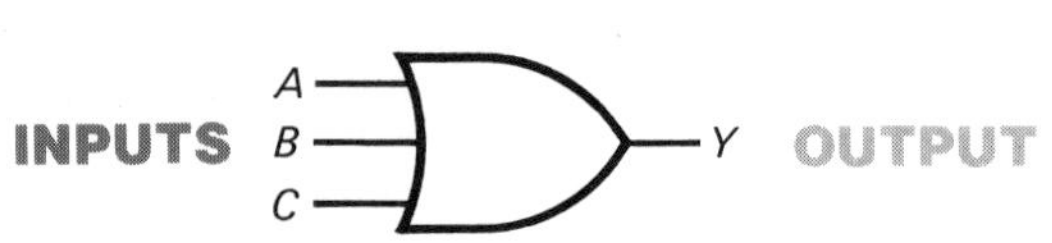

Now suppose you are given the Boolean expression $\overline{A} \cdot B + A \cdot \overline{B} + \overline{B} \cdot C = Y$ (read as "not A and B, or A and not B, or not B and C equals output Y"). How would you construct a circuit that would do the job of this expression? The first step is to look at the Boolean expression and note that you must OR $\overline{A} \cdot B$ with $A \cdot \overline{B}$ with $\overline{B} \cdot C$. Figure 4-2(*a*) shows that a 3-input OR gate will form the output Y. This may be redrawn as in Figure 4-2(*b*).

INPUTS B — Y OUTPUT

Fig. 4-1 Logic diagram for Boolean expression $A + B + C = Y$.

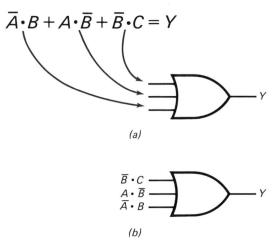

$$\overline{A} \cdot B + A \cdot \overline{B} + \overline{B} \cdot C = Y$$

(a)

(b)

Fig. 4-2 Step 1 in constructing a logic circuit.

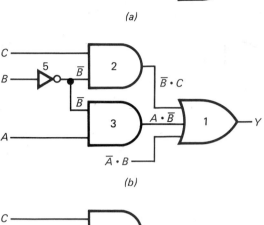

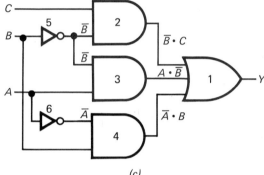

Fig. 4-3   Step 2 in constructing a logic circuit.

The second step in constructing a logic circuit from the given Boolean expression $\overline{A} \cdot B + A \cdot \overline{B} + \overline{B} \cdot C = Y$ is shown in Fig. 4-3. Notice in (a) that an AND gate has been added to feed the $\overline{B} \cdot C$ to the OR gate and an inverter has been added to form the $\overline{B}$ for the input to AND gate 2. Figure 4-3(b) adds AND gate 3 to form the $A \cdot \overline{B}$ input to the OR gate. Finally, Fig. 4-3(c) adds AND gate 4 and inverter 6 to form the $\overline{A} \cdot B$ input to the OR gate. Figure 4-3(c) is the circuit that would be constructed to perform the required logic given in the Boolean expression $\overline{A} \cdot B + A \cdot \overline{B} + \overline{B} \cdot C = Y$.

Notice that we started at the output of the logic circuit and worked toward the inputs. You have now experienced how combinational logic circuits are constructed from Boolean expressions.

Boolean expressions come in two forms. The *sum-of-products* form is the type we saw in Fig. 4-2. Another example of this form is $A \cdot B + B \cdot C = Y$. The other Boolean expression form is the *product-of-sums*; an ex-

ample $(D + E) \cdot (E + F) = Y$. The sum-of-products form is called the *minterm form* in engineering texts. The product-of-sums form is called the *maxterm form* by engineers, technicians, and scientists.

4-2 DRAWING A CIRCUIT FROM A MAXTERM BOOLEAN EXPRESSION

Suppose you are given the maxterm Boolean expression $(A + B + C) \cdot (\overline{A} + \overline{B}) = Y$. The first step in constructing a logic circuit for this Boolean expression is shown in Fig. 4-4(a). Notice that the terms $(A + B + C)$ and $(\overline{A} + \overline{B})$ must be ANDed together to form output Y. Figure 4-4(b) shows the logic circuit redrawn. The second step in drawing the logic circuit is shown in Fig. 4-5. The $(\overline{A} + \overline{B})$ part of the expression is produced by adding OR gate 2 and inverters 3 and 4, as illustrated in Fig. 4-5(a). Then, the expression $(A + B + C)$ is delivered to the AND gate by OR gate 5 in Fig. 4-5(b). The logic circuit shown in Fig. 4-5(b) is the complete logic circuit for the maxterm Boolean expression $(A + B + C) \cdot (\overline{A} + \overline{B}) = Y$.

In summary, we work from right to left (from output to input) when converting a Boolean expression to a logic circuit. Notice that we only use AND, OR, and NOT gates when constructing combinational logic circuits. Maxterm and minterm Boolean expressions both can be converted to logic circuits.

You now should be able to identify minterm and maxterm Boolean expressions. And you should be able to convert Boolean expressions

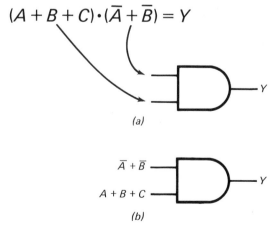

Fig. 4-4   Step 1 in constructing a product-of-sums logic circuit.

From page 24: Combinational logic circuits

On this page: Sum-of-products

Product-of-sums

Minterm form

Maxterm form

25

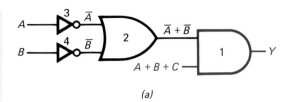

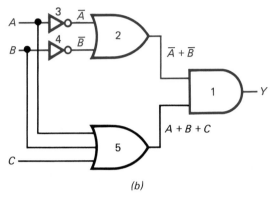

(b)

Fig. 4-5 Step 2 in constructing a product-of-sums logic circuit.

into a combinational logic circuits by using AND, OR, and NOT gates.

4-3 TRUTH TABLES AND BOOLEAN EXPRESSIONS

Boolean expressions are a convenient method of describing how a logic circuit will operate. The *truth table* is another precise method of describing how a logic circuit will work. As you work in digital electronics you will have to convert information from truth-table form to a Boolean expression.

Look at the truth table in Fig. 4-6(a). Notice that only two of the eight possible combinations of inputs A, B, and C will generate a logical 1 at the output. The two combinations that generate a 1 output are shown as $\overline{C} \cdot B \cdot A$ (read as "not C and B and A") *or* $C \cdot \overline{B} \cdot \overline{A}$ (read as "C and not B and not A"). Fig. 4-6(b) shows how the two combinations are ORed together to form the Boolean expression for the truth table. Both the truth table in Fig. 4-6(a) and the Boolean expression in Fig. 4-6(b) describe how the logic circuit should work.

The truth table is the beginning of most logic circuits. You must be able to convert the truth-table information into a Boolean expression as in this section. Remember to look for the variables that generate a logical 1 in the truth table.

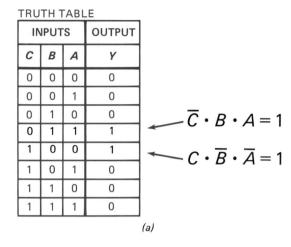

(a)

(b) Boolean expression

$$\overline{C} \cdot B \cdot A + C \cdot \overline{B} \cdot \overline{A} = Y$$

Fig. 4-6 Forming a Boolean expression from truth table.

Occasionally you must reverse the procedure you have just learned. That is, you must take a Boolean expression and from it construct a truth table. Consider the Boolean expression in Fig. 4-7(a). It appears that two combinations of inputs A, B, and C will generate a logical 1 at the output. In Fig. 4-7(b) we find the correct combinations of A, B, and C that are given in the Boolean expression and mark a 1 in the output column. All other outputs in the truth table are 0. Both the Boolean expression in Fig. 4-7(a) and the

(a) Boolean expression

$$\overline{C} \cdot B \cdot \overline{A} + C \cdot \overline{B} \cdot A = Y$$

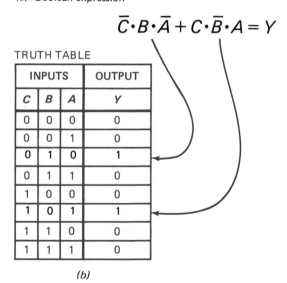

(b)

Fig. 4-7 Constructing a truth table from a Boolean expression.

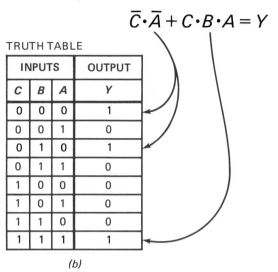

(a) Boolean expression

$$\overline{C}\cdot\overline{A} + C\cdot B\cdot A = Y$$

TRUTH TABLE

INPUTS			OUTPUT
C	B	A	Y
0	0	0	1
0	0	1	0
0	1	0	1
0	1	1	0
1	0	0	0
1	0	1	0
1	1	0	0
1	1	1	1

(b)

Fig. 4-8 Constructing a truth table from a Boolean expression.

truth table in Fig. 4-7(b) accurately describe the operation of the same logic circuit.

Suppose you are given the Boolean expression in Fig. 4-8(a). At first glance it seems that this would produce two outputs with a logical 1. However, if you look closely at Fig. 4-8(b) you will see that the Boolean expression $\overline{C}\cdot\overline{A} + C\cdot B\cdot A = Y$ actually generates three logical 1s in the output column. The "trick" illustrated in Fig. 4-8 should make you very cautious. Make sure you have all the combinations that will generate a logical 1 in the truth table. The Boolean expression in Fig. 4-8(a) and the truth table in Fig. 4-8(b) both describe the same logic circuit.

You have now converted truth tables into Boolean expressions and Boolean expressions into truth tables. You were reminded that the Boolean expressions you worked with were minterm Boolean expressions. The procedure for producing maxterm Boolean expressions from a truth table is quite different.

4-4 A SAMPLE PROBLEM

The procedures in Secs. 4-1 to 4-3 are needed skills as you work in digital electronics. To assist you we shall take an everyday logic problem and work from truth table to Boolean expression to logic circuit.

Let us assume that we are designing a simple *electronic lock*. The lock will open only when certain switches are pressed. Figure 4-9(a) is the truth table for the electronic

Electronic lock

TRUTH TABLE

INPUTS			OUTPUT
C	B	A	Y
0	0	0	0
0	0	1	0
0	1	0	0
0	1	1	0
1	0	0	1
1	0	1	0
1	1	0	0
1	1	1	1

(a)

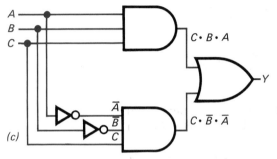

(b) $$C\cdot B\cdot A + C\cdot\overline{B}\cdot\overline{A} = Y$$

(c)

Fig. 4-9 Electronic lock problem. (a) Truth table. (b) Boolean expression. (c) Logic circuit.

lock. Notice that two combinations of input switches A, B, and C will generate a 1 at the output. A 1 output will open the lock. Figure 4-9(b) shows how we form the minterm Boolean expression for the electronic lock circuit. The logic circuit in Figure 4-9(c) is then drawn from the Boolean expression. Look over the sample problem in Fig. 4-9 and be sure you can follow how we converted from the truth table to the Boolean expression and then to the logic circuit.

You now should be able to solve a problem such as the one in Fig. 4-9. The following self test will give you some practice in solving problems dealing with truth tables, Boolean expressions, and combinational logic circuits.

Self Test

Check your understanding by answering questions 1 to 6.

1. Construct a logic circuit using AND, OR, and NOT gates from the following Boolean expressions:

27

a. $\overline{A} \cdot \overline{B} + A \cdot B = Y$
b. $(A + B) \cdot (\overline{A} + \overline{B}) = Y$
c. $\overline{A} \cdot \overline{C} + A \cdot B \cdot C = Y$
d. $(\overline{A} + B) \cdot \overline{C} = Y$

2. Refer to question 1. Determine whether the Boolean expression *a* to *d* are minterm or maxterm expressions.

3. Construct a truth table for each of the following Boolean expressions (use *B* and *A* for two variables and *C*, *B*, and *A* for three variables):

a. $\overline{A} \cdot \overline{B} + A \cdot B = Y$
b. $A \cdot \overline{B} + \overline{A} \cdot B = Y$
c. $\overline{A} \cdot B \cdot C + A \cdot \overline{B} \cdot C = Y$
d. $\overline{A} \cdot \overline{C} + A \cdot B \cdot C = Y$

4. A minterm Boolean expression is also called a ___?___ (product-of-sums, sum-of-products) form expression.

5. A maxterm Boolean expression is also called a ___?___ (product-of-sums, sum-of-products) form expression.

6. Given the following truth table for an electronic lock, first write the minterm Boolean expression and then draw the logic circuit for the lock (use AND, OR, and NOT gates):

TRUTH TABLE

INPUT SWITCHES			OUTPUT
C	B	A	Y
0	0	0	0
0	0	1	0
0	1	0	1
0	1	1	0
1	0	0	0
1	0	1	1
1	1	0	0
1	1	1	0

4-5 SIMPLIFYING BOOLEAN EXPRESSIONS

Consider the Boolean expression $\overline{A} \cdot B + A \cdot \overline{B} + A \cdot B = Y$ in Fig. 4-10(*a*). In constructing a logic circuit for this Boolean expression, we would find that we would need three AND gates, two inverters, and one 3-input OR gate. Figure 4-10(*b*) is a logic circuit that would perform the logic of the Boolean expression $\overline{A} \cdot B + A \cdot \overline{B} + A \cdot B = Y$. Fig-

(a) Original Boolean expression

$$\overline{A} \cdot B + A \cdot \overline{B} + A \cdot B = Y$$

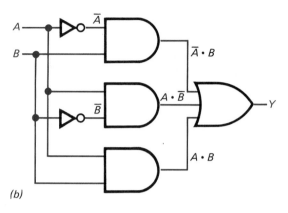

(b)

(c)

TRUTH TABLE

INPUTS		OUTPUT
B	A	Y
0	0	0
0	1	1
1	0	1
1	1	1

(d) Simplified Boolean expression

$$A + B = Y$$

(e)

Fig. 4-10 Simplifying Boolean expressions.

ure 4-10(*c*) diagrams the truth table for the Boolean expression and logic circuit in Fig. 4-10(*a*) and (*b*). Immediately you recognize the truth table in Fig. 4-10(*c*) as the truth table for a 2-input OR gate. The simple Boolean expression for a 2-input OR gate is $A + B = Y$, as shown in Fig. 4-10(*d*). The logic circuit for a 2-input OR gate in its simplest form is diagramed in Fig. 4-10(*e*).

The example summarized in Fig. 4-10 shows how we must try to simplify our original Boolean expression to get a simple, inexpensive logic circuit. In this case we were lucky enough to notice that the truth table belonged to an OR gate. However, usually we must use more systematic methods of simplifying our Boolean expression. Such methods include applying Boolean algebra and *Karnaugh mapping.*

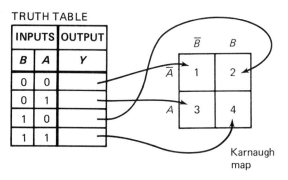

Fig. 4-13 Looping 1s together on a Karnaugh map.

Fig. 4-11 The meaning of squares in a Karnaugh map.

Boolean algebra was originated by George Boole (1815–1864). Boole's algebra was adapted in the 1930s for use in digital logic circuits; it is the basis for the tricks we shall use to simplify Boolean expressions. We shall not deal directly with Boolean algebra in this text. Many of you who continue on in digital electronics and engineering will study Boolean algebra in detail.

Karnaugh mapping, an easy-to-use method of simplifying Boolean expressions, is covered in detail in Secs. 4-6 to 4-9. Several other simplification methods are available, including Veitch diagrams, Venn diagrams, and the tabular method of simplification.

4-6 KARNAUGH MAPS

In 1953 Maurice Karnaugh published an article about his system of mapping and thus simplifying Boolean expressions. Figure 4-11 illustrates a Karnaugh map. The four squares (1, 2, 3, 4) represent the four possible combinations of A and B in a two-variable truth table. Square 1 in the Karnaugh map, then, stands for $\overline{A} \cdot \overline{B}$, square 2 for $\overline{A} \cdot B$, and so forth.

Let us map the familiar problem from Fig. 4-10. The original Boolean expression

$\overline{A} \cdot B + A \cdot \overline{B} + A \cdot B = Y$ is rewritten in Fig. 4-12(a) for your convenience. Next, a 1 is placed in each square of the Karnaugh map represented in the original Boolean expression, as shown in Fig. 4-12(b). The filled-in Karnaugh map is now ready for *looping*. The looping technique is shown in Fig. 4-13. *Adjacent 1s* are *looped together* in groups of two, four, or eight. Looping continues until all 1s are included inside a loop. Each loop is a new term in the simplified Boolean expression. Notice that we have two loops in Fig. 4-13. These two loops mean that we shall have two terms ORed together in our new simplified Boolean expression.

Now let us simplify the Boolean expression based upon the two loops that are redrawn in Fig. 4-14. First the bottom loop: notice that A is included along with a B and a $\overline{B}$. The B and $\overline{B}$ terms can be *eliminated* according to rules of Boolean algebra. This leaves the term A in the bottom loop. Likewise, the vertical loop contains an A and a $\overline{A}$, which are eliminated, leaving only term B. The leftover A and B terms are then ORed together, giving the simplified Boolean expression $A + B = Y$.

The procedure for simplifying a Boolean expression sounds complicated. Actually, this procedure is quite easy after some practice. Here is a summary of the six steps:

1. Start with a minterm Boolean expression.
2. Record 1s on a Karnaugh map.

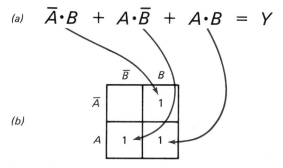

Fig. 4-12 Marking 1s on a Karnaugh map.

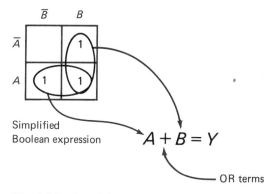

Fig. 4-14 Simplifying a Boolean expression from a Karnaugh map.

29

3. Loop adjacent ls (loops of two, four, or eight squares).
4. Simplify by dropping terms that contain a term and its complement within a loop.
5. OR the remaining terms (one term per loop).
6. Write the simplified minterm Boolean expression.

4-7 KARNAUGH MAPS WITH THREE VARIABLES

Consider the unsimplified Boolean expression $A \cdot \overline{B} \cdot \overline{C} + \overline{A} \cdot \overline{B} \cdot \overline{C} + \overline{A} \cdot \overline{B} \cdot C + A \cdot B \cdot \overline{C} = Y$, as given in Fig. 4-15(a). A three-variable Karnaugh map is illustrated in Fig. 4-15(b). Notice the eight possible combinations of A, B, and C, which are represented by the eight squares in the map. Tabulated on the map are four 1s, which represent each of the four terms in the original Boolean expression. The filled-in Karnaugh map is redrawn in Fig. 4-15(c). Adjacent groups of two 1s are looped. The bottom loop contains both B and $\overline{B}$. The B and $\overline{B}$ terms are eliminated. The bottom loop still contains the A and $\overline{C}$, giving the term $A \cdot \overline{C}$. The upper loop contains both a C and a $\overline{C}$. The C and $\overline{C}$ terms are eliminated, leaving the $\overline{A} \cdot \overline{B}$ term. A minterm Boolean expression is formed by adding the OR symbol. The simplified Boolean expression is written in Fig. 4-15(d) as $A \cdot \overline{C} + \overline{A} \cdot \overline{B} = Y$.

You can see that the simplified Boolean expression in Fig. 4-15 would take fewer electronic parts than the original expression. Remember that the much different looking simplified Boolean expression will produce the same truth table as the original Boolean expression.

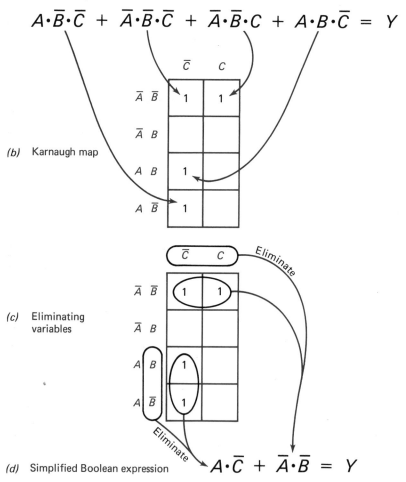

(a) Boolean expression

$$A \cdot \overline{B} \cdot \overline{C} + \overline{A} \cdot \overline{B} \cdot \overline{C} + \overline{A} \cdot \overline{B} \cdot C + A \cdot B \cdot \overline{C} = Y$$

(b) Karnaugh map

(c) Eliminating variables

(d) Simplified Boolean expression $A \cdot \overline{C} + \overline{A} \cdot \overline{B} = Y$

Fig. 4-15 Simplifying a Boolean expression using a Karnaugh map. (*a*) Unsimplified expression. (*b*) Mapping 1s. (*c*) Looping 1s. (*d*) Forming simplified minterm expression.

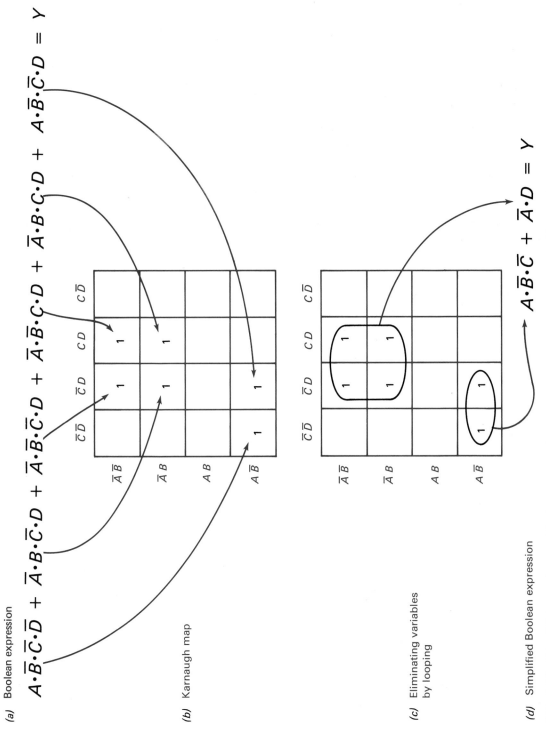

(a) Boolean expression

$$A \cdot \overline{B} \cdot \overline{C} \cdot \overline{D} + \overline{A} \cdot B \cdot \overline{C} \cdot D + \overline{A} \cdot \overline{B} \cdot \overline{C} \cdot D + \overline{A} \cdot B \cdot C \cdot D + \overline{A} \cdot \overline{B} \cdot C \cdot D + A \cdot \overline{B} \cdot \overline{C} \cdot D = Y$$

(b) Karnaugh map

(c) Eliminating variables by looping

(d) Simplified Boolean expression

$$A \cdot \overline{B} \cdot \overline{C} + \overline{A} \cdot D = Y$$

Fig. 4-16 Simplifying a six-term Boolean expression to a two-term expression using a Karnaugh map.

4-8 KARNAUGH MAPS WITH FOUR VARIABLES

The truth table for four variables would have 16 possible combinations. Therefore simplifying a Boolean expression that has four variables sounds complicated. But a Karnaugh map makes the job of simplifying a Boolean expression easy.

Consider the Boolean expression $A \cdot \bar{B} \cdot \bar{C} \cdot \bar{D} + \bar{A} \cdot B \cdot \bar{C} \cdot D + \bar{A} \cdot \bar{B} \cdot \bar{C} \cdot D + \bar{A} \cdot \bar{B} \cdot C \cdot D + \bar{A} \cdot B \cdot C \cdot D + A \cdot \bar{B} \cdot \bar{C} \cdot D = Y$, as in Fig. 4-16(a) on page 31. The four-variable Karnaugh map in Fig. 4-16(b) gives the 16 possible combinations of A, B, C, and D. These are represented in the 16 squares of the map. Tabulated on the map are six 1s, which represent the six terms in the original Boolean expression. The Karnaugh map is redrawn in Fig. 4-16(c). Adjacent groups of two 1s and four 1s are looped. The bottom loop of two 1s eliminates the D and $\bar{D}$ terms. The bottom loop then produces the $A \cdot \bar{B} \cdot \bar{C}$ term. The upper loop of four 1s eliminates the C and $\bar{C}$ and B and $\bar{B}$ terms. The upper loop then produces the $\bar{A} \cdot D$ term. The $A \cdot \bar{B} \cdot \bar{C}$ and $\bar{A} \cdot D$ terms are then ORed together. The simplified minterm Boolean expression is written in Fig. 4-16(d) as $A \cdot \bar{B} \cdot \bar{C} + \bar{A} \cdot D = Y$.

Observe that the same procedure and rules are used for simplifying Boolean expressions with two, three, or four variables and that larger loops in a Karnaugh map eliminate more variables. You must take care in making sure the maps look just like the ones in Figs. 4-14 to 4-16.

4-9 MORE KARNAUGH MAPS

This section presents some sample Karnaugh maps. Notice the unusual looping procedures used on most maps in this section.

Consider the Boolean expression in Fig. 4-17(a). The four terms are shown as four 1s on the Karnaugh map in Fig. 4-17(b). The correct looping procedure is shown. Notice that the Karnaugh map is considered wrapped in a cylinder, with the left side adjacent to the right side. Also notice the elimination of the A and $\bar{A}$ and C and $\bar{C}$ terms. The simplified Boolean expression of $B \cdot \bar{D} = Y$ is shown in Fig. 4-17(c).

Another unusual looping variation is illustrated in Fig. 4-18(a). Notice that the top and

(a) Boolean expression

$$A \cdot B \cdot \bar{C} \cdot \bar{D} + \bar{A} \cdot B \cdot \bar{C} \cdot \bar{D} +$$
$$\bar{A} \cdot B \cdot C \cdot \bar{D} + A \cdot B \cdot C \cdot \bar{D} = Y$$

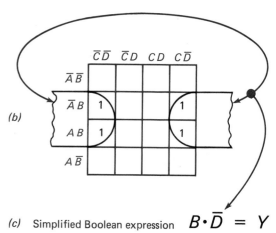

(c) Simplified Boolean expression $B \cdot \bar{D} = Y$

Fig. 4-17 Simplifying a Boolean expression using a Karnaugh map. By considering the map as a vertical cylinder, the four 1s can be looped as shown.

bottom of the map are adjacent to one another, as if rolled into a cylinder while looping. The simplified Boolean expression for this map is given as $\bar{B} \cdot \bar{C} = Y$ in Fig. 4-18(b).

Figure 4-19(a) shows still another unusual looping pattern. The four corners of the Karnaugh map are considered connected, as if the map were formed into a ball. The four corners are then adjacent and may be formed into one loop as shown. The simplified Boolean expression is the term $\bar{B} \cdot \bar{D} = Y$, given in Fig. 4-19(b).

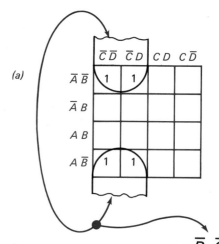

(b)   Simplified Boolean expression $\bar{B} \cdot \bar{C} = Y$

Fig. 4-18 Simplifying a Boolean expression by considering the Karnaugh map as a horizontal cylinder. In this way, the four 1s can be looped.

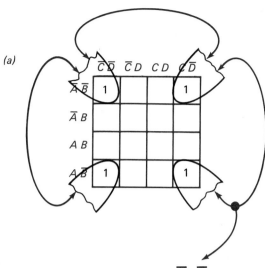

(a)

(b) Simplified Boolean expression $\overline{B} \cdot \overline{D} = Y$

Fig. 4-19 Simplifying a Boolean expression by thinking of the Karnaugh map as a ball. In this way, the 1s in the four corners can be looped.

Self Test

Check your understanding by solving Probs. 7 to 10.

7. Simplify the Boolean expression
$$\overline{A} \cdot B \cdot \overline{C} + \overline{A} \cdot B \cdot C + A \cdot B \cdot \overline{C} + A \cdot B \cdot C = Y$$
by
a. Plotting 1s on a three-variable Karnaugh map
b. Looping groups of two or four 1s
c. Eliminating variables whose complements appear within the loop(s)
d. Writing the simplified Boolean expression

8. Simplify the Boolean expression
$$\overline{A} \cdot B \cdot \overline{C} \cdot \overline{D} + A \cdot B \cdot \overline{C} \cdot \overline{D} + \overline{A} \cdot B \cdot \overline{C} \cdot D + A \cdot B \cdot \overline{C} \cdot D + A \cdot \overline{B} \cdot C \cdot D + A \cdot \overline{B} \cdot C \cdot \overline{D} = Y$$
by
a. Plotting 1s on a 4-variable Karnaugh map
b. Looping groups of two or four 1s
c. Eliminating variables whose complements appear within loops
d. Writing the simplified Boolean expression

9. Simplify the Boolean expression
$$\overline{A} \cdot B \cdot \overline{C} \cdot \overline{D} + \overline{A} \cdot B \cdot \overline{C} \cdot D + \overline{A} \cdot \overline{B} \cdot C \cdot D + \overline{A} \cdot B \cdot C \cdot \overline{D} + A \cdot \overline{B} \cdot \overline{C} \cdot D + A \cdot \overline{B} \cdot C \cdot D = Y$$

by
a. Plotting 1s on a four-variable Karnaugh map
b. Looping groups of two or four 1s
c. Eliminating variables whose complements appear within loops
d. Writing the simplified Boolean expression

10. Simplify the Boolean expression
$$\overline{A} \cdot \overline{B} \cdot \overline{C} + \overline{A} \cdot \overline{B} \cdot C + A \cdot \overline{B} \cdot \overline{C} + A \cdot \overline{B} \cdot C + A \cdot B \cdot C = Y$$
by
a. Plotting 1s on a three-variable Karnaugh map
b. Looping groups of two or four 1s
c. Eliminating variables whose complements appear within loops
d. Writing the simplified Boolean expression

4-10 USING NAND LOGIC

In Chap. 3, Sec. 3-8 explained how NAND gates could be wired to form other gates or inverters (see Fig. 3-20). We mentioned that the NAND gate can be used as a universal gate. In this section, you shall see how NAND gates are used in wiring combinational logic circuits. NAND gates are widely employed in industry because they are easy to use and readily available.

Suppose your supervisor gives you the Boolean expression $A \cdot B + A \cdot \overline{C} = Y$, as shown in Fig. 4-20(*a*). You are told to solve this logic problem at the least cost. You first draw the logic circuit for the Boolean expression shown in Fig. 4-20(*b*), using AND gates, an OR gate, and an inverter. Checking a manufacturer's catalog, you determine that you must use three different ICs to do the job.

Your supervisor suggests that you try using NAND logic. You redraw your logic circuit to look like the NAND gate circuit in Fig. 4-20(*c*). Upon checking a catalog, you find you need only one IC that contains the four NAND gates to do the job. Recall from Chap. 3 that the OR symbol with invert bubbles at the inputs is another symbol for a NAND gate. You finally test the circuit in Fig. 4-20(*c*) and find that it performs the logic $A \cdot B + A \cdot \overline{C} = Y$. Your supervisor is pleased you have found a circuit that requires only one IC, as compared to the circuit in Fig. 4-20(*b*), which uses three ICs.

Remembering this trick will help you appre-

Universal gate

Using NAND logic

33

(a) $A \cdot B + A \cdot \overline{C} = Y$

Data selectors

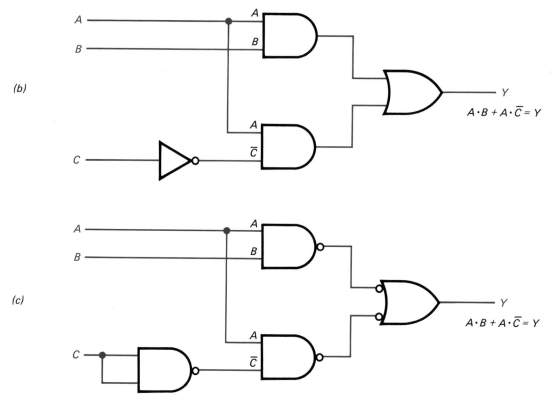

(b)

(c)

Fig. 4-20 Using NAND gates in logic circuits. (*a*)Boolean expression. (*b*) AND-OR logic circuit. (*c*) Equivalent NAND logic circuit.

ciate *why* NAND gates are used in many logic circuits. If your future job is in digital circuit design, this can be a useful tool for making your final circuit the best for the least cost.

You may have questioned why the NAND gates in Fig. 4-20(*c*) could be substituted for the AND and OR gates in Fig. 4-20(*b*). If you look carefully at Fig. 4-20(*c*), you will see two AND symbols feeding into an OR symbol. There are two invert bubbles between the output of the AND symbols and the inputs of the OR symbol. From previous experience we know that if we invert twice we have the original logic state. Hence the two invert bubbles in Fig. 4-20(*c*) between the AND and OR symbols cancel one another. Because the two invert bubbles cancel one another, we end up with two AND gates feeding an OR gate.

4-11 SOLVING LOGIC PROBLEMS THE EASY WAY

Manufacturers of ICs have simplified the job of solving many combinational logic problems

by producing *data selectors*. A data selector is often a *one-package solution* to a complicated logic problem. The data selector actually contains a rather large number of gates packaged inside a single IC. The data selector will be used as a "universal package" for solving combinational logic problems.

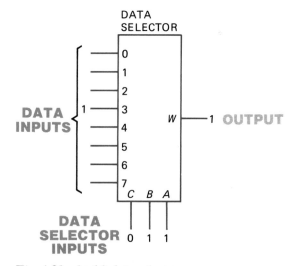

Fig. 4-21 1-of-8 data selector.

A *1-of-8 data selector* is illustrated in Fig. 4-21. Notice the eight *data inputs* numbered from 0 to 7 on the left. Also notice the three *data selector inputs* labeled A, B, and C at the bottom of the data selector. The output of the data selector is labeled W.

The basic job the data selector performs is transferring data from a *given* data input (0 to 7) to the output (W). The selection of which data input will be selected is determined by which binary number you place on the data selector inputs at the bottom (see Fig. 4-21). The data selector in Fig. 4-21 functions in the same manner as a rotary switch. Figure 4-22 shows the data at input 3 being transferred to the output by the rotary switch contacts. In like manner the data from data input 3 in Fig.

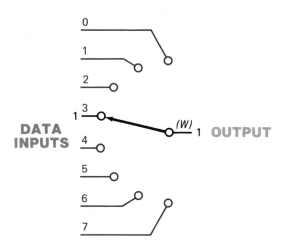

Fig. 4-22 One-pole eight-position rotary switch works as a data selector.

1-of-8 data selector

(a) Simplified Boolean expression

$$A \cdot B \cdot C \cdot D + \bar{A} \cdot \bar{B} \cdot \bar{C} \cdot \bar{D} + A \cdot \bar{B} \cdot \bar{C} \cdot D + A \cdot B \cdot \bar{C} \cdot \bar{D} +$$
$$\bar{A} \cdot B \cdot C \cdot \bar{D} + \bar{A} \cdot B \cdot \bar{C} \cdot D + \bar{A} \cdot \bar{B} \cdot C \cdot D = Y$$

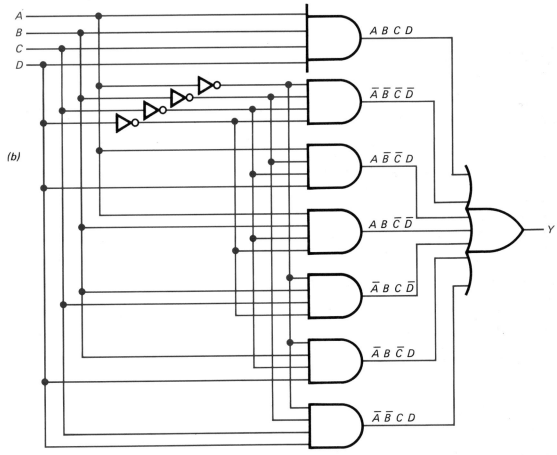

Fig. 4-23 *(a)* Simplified Boolean expression. *(b)* Logic circuit for Boolean expression.

4-21 is being transferred to output W of the data selector. In the rotary switch you must mechanically change the switch position to transfer data from another input. In the 1-of-8 data selector in Fig. 4-21 we need only change the binary input at the data selector inputs to transfer data from another data input to the output. Remember that the data selector operates somewhat as a rotary switch in transferring logical 0s or 1s from a given input to the single output.

1-of-16 data selector

Now you will learn how data selectors can be used to solve logic problems. Consider the *simplified* Boolean expression shown in Fig. 4-23(*a*). For your convenience a logic circuit for this complicated Boolean expression is drawn in Fig. 4-23(*b*). Using standard ICs, we probably would have to use from six to nine ICs to solve this problem. This would be quite expensive in regards to the cost of the ICs and printed circuit board space.

A less costly solution to solving the logic problem is to use a data selector. The Boolean expression from Fig. 4-23(*a*) is repeated in truth-table form in Fig. 4-24(*a*). A *1-of-16 data selector* is added to the truth table in Fig. 4-24(*b*). Notice that logical 0s and 1s are placed at the 16 data inputs of the data selector corresponding to the truth-table output column Y. These are *permanently* connected for this truth table. Data selector inputs (*D*, *C*, *B*, and A) are switched to the binary numbers on the input side of the truth table. If the data selector inputs *D*, *C*, *B*, and A are at binary 0000, then a logical 1 is transferred to output W of the data selector. The first line of the truth table requires that a logical 1 appear at output W when *D*, *C*, *B*, and A are all 0s. If data selector inputs *D*, *C*, *B*, and A are at binary 0001, a logical 0 appears at output W as required by the truth table. Any combination of *D*, *C*, *B*, and A will generate the proper output according to the truth table.

We used the data selector to solve a complicated logic problem. In Fig. 4-23 we found

(a)

TRUTH TABLE

INPUTS				OUTPUT
D	*C*	*B*	*A*	*Y*
0	0	0	0	1
0	0	0	1	0
0	0	1	0	0
0	0	1	1	1
0	1	0	0	0
0	1	0	1	0
0	1	1	0	1
0	1	1	1	0
1	0	0	0	0
1	0	0	1	1
1	0	1	0	1
1	0	1	1	0
1	1	0	0	1
1	1	0	1	0
1	1	1	0	0
1	1	1	1	1

(b)

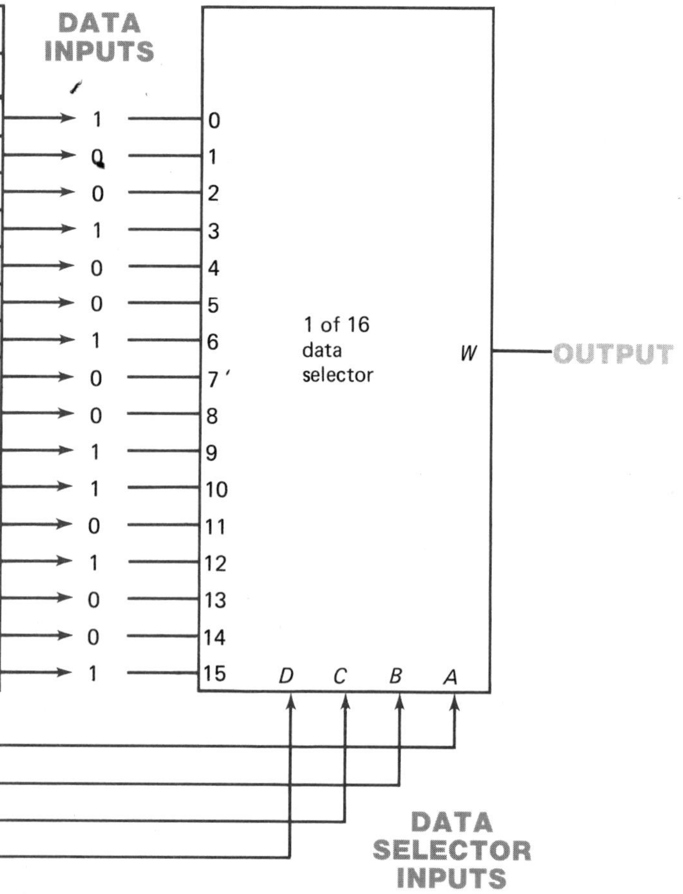

Fig. 4-24 Solving logic problem with a data selector.

we needed at least six ICs to solve this logic problem. Using the data selector in Fig. 4-24, we solved this problem by using only one IC.

The data selector seems to be an easy-to-use and efficient way to solve combinational logic problems. Commonly available data selectors can solve logic problems with three, four, or five variables. When using manufacturers' data manuals you will notice that data selectors are also called *multiplexers*.

Multiplexers

Summary

1. Combining gates in combinational logic circuits from Boolean expressions is a necessary skill for most competent technicians and engineers.

2. Workers in digital electronics must have an excellent knowledge of gate symbols, truth tables, and Boolean expressions and know how to convert from one form to another.

(a) Minterm Boolean expression

$$A \cdot B + \overline{A} \cdot \overline{C} = Y$$

(c) Maxterm Boolean expression

$$(A + \overline{C}) \cdot (\overline{A} + B) = Y$$

(b)

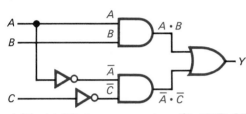

(d)

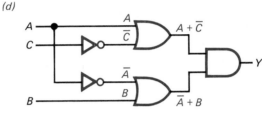

Fig. 4-25 *(a)* Minterm expression. *(b)* AND-OR logic circuit. *(c)* Maxterm expression. *(d)* OR-AND logic circuit.

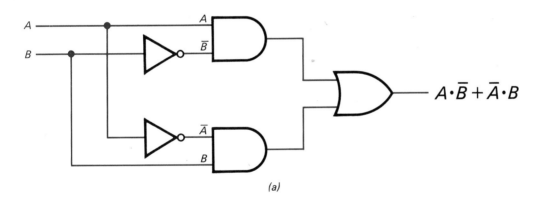

(a)

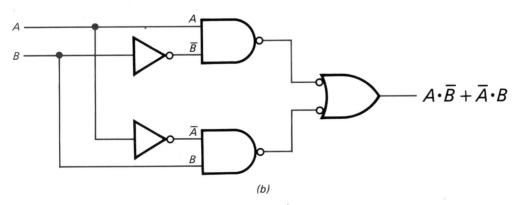

(b)

Fig. 4-26 *(a)* AND-OR logic circuit. *(b)* Equivalent NAND logic circuit.

37

3. The minterm Boolean expression (sum-of-products form) might look like the expression in Fig. 4-25(a). The Boolean expression $A \cdot B + \overline{A} \cdot \overline{C} = Y$ would be wired as shown in Fig. 4-25(b).

4. The maxterm Boolean expression (product-of-sums form) might look like the expression in Fig. 4-25(c). The Boolean expression $(A + \overline{C}) \cdot (\overline{A} + B) = Y$ would be wired as shown in Fig. 4-25(d).

5. A Karnaugh map is a convenient method of simplifying Boolean expressions.

6. AND-OR logic circuits can be wired easily by using only NAND gates, as shown in Fig. 4-26.

7. Data selectors are a simple one-package method of solving many gating problems

Questions

1. Engineers and technicians refer to circuits that are a combination of different gates as a ____?____ logic circuit.

2. Draw a logic diagram for the Boolean expression $\overline{A} \cdot \overline{B} + B \cdot C = Y$. Use one OR gate, two AND gates, and two inverters.

3. The Boolean expression $\overline{A} \cdot \overline{B} + B \cdot C = Y$ is in ____?____ (product-of-sums, sum-of-products) form.

4. The Boolean expression $(A + B) \cdot (C + D) = Y$ is in ____?____ (product-of-sums, sum-of-products) form.

5. A Boolean expression in product-of-sums form is also called a ____?____ expression.

6. A Boolean expression in sum-of-products form is also called a ____?____ expression.

7. Write the minterm Boolean expression that would describe the truth table in Fig. 4-27. Do not simplify the Boolean expression.

8. Draw a truth table (three variable) that represents the Boolean expression $\overline{C} \cdot \overline{B} + C \cdot \overline{B} \cdot A = Y$.

9. The truth table in Fig. 4-28 is for an electronic lock. The lock will open only when a logical 1 appears at the output. First, write the minterm Boolean expression for the lock. Second, draw the logic circuit for the lock (use AND, OR, and NOT gates).

10. List the six steps for simplifying a Boolean expression as discussed in Sec. 4-6.

TRUTH TABLE

INPUTS			OUTPUT
C	B	A	Y
0	0	0	1
0	0	1	0
0	1	0	1
0	1	1	0
1	0	0	0
1	0	1	1
1	1	0	0
1	1	1	1

Fig. 4-27   Truth table.

TRUTH TABLE

INPUTS			OUTPUT
C	B	A	Y
0	0	0	0
0	0	1	0
0	1	0	0
0	1	1	1
1	0	0	1
1	0	1	0
1	1	0	0
1	1	1	0

Fig. 4-28   Truth table for electronic lock.

11. Use a Karnaugh map to simplify the Boolean expression

$$\overline{A} \cdot \overline{B} \cdot \overline{C} + \overline{A} \cdot \overline{B} \cdot C + A \cdot B \cdot \overline{C} + A \cdot \overline{B} \cdot \overline{C} = Y$$

Write the simplified Boolean expression in minterm form.

12. Use a Karnaugh map to simplify the Boolean expression

$$A \cdot \overline{B} \cdot \overline{C} \cdot \overline{D} + A \cdot \overline{B} \cdot \overline{C} \cdot D + A \cdot \overline{B} \cdot C \cdot D + A \cdot \overline{B} \cdot C \cdot \overline{D} = Y$$

13. From the truth table in Fig. 4-27 do the following:
    a. Write the unsimplified Boolean expression.
    b. Use a Karnaugh map to simplify the Boolean expression from (*a*).
    c. Write the simplified minterm Boolean expression for the truth table.
    d. Draw a logic circuit from the simplified Boolean expression (use AND, OR, and NOT gates).
    e. Redraw the logic circuit from *d* using only NAND gates.

14. Use a Karnaugh map to simplify the Boolean expression

$$\overline{A} \cdot \overline{B} \cdot C \cdot D + A \cdot B \cdot \overline{C} \cdot \overline{D} + A \cdot B \cdot C \cdot \overline{D} + A \cdot \overline{B} \cdot C \cdot D = Y$$

Write the answer as a minterm Boolean expression.

15. From the Boolean expression

$$\overline{A} \cdot \overline{B} \cdot \overline{C} \cdot \overline{D} + \overline{A} \cdot \overline{B} \cdot C \cdot D + \overline{A} \cdot B \cdot \overline{C} \cdot D + A \cdot B \cdot C \cdot D$$
$$+ A \cdot B \cdot C \cdot \overline{D} + A \cdot \overline{B} \cdot \overline{C} \cdot \overline{D} = Y$$

do the following:
    a. Draw a truth table for this expression.
    b. Use a Karnaugh map to simplify.
    c. Draw a logic circuit of the simplified Boolean expression (use AND, OR, and NOT gates).
    d. Draw a circuit to solve this problem using a 1-of-16 data selector.

Answers to Self Tests

1. *a.*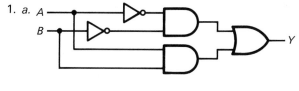

b.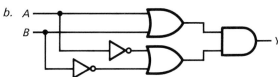

c.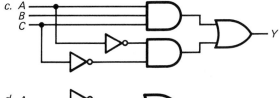

d.

2. *a.* Minterm expression
   b. Maxterm expression
   c. Minterm expression
   d. Maxterm expression

3. *a.*

B	A	Y
0	0	1
0	1	0
1	0	0
1	1	1

b.

B	A	Y
0	0	0
0	1	1
1	0	1
1	1	0

c.

C	B	A	Y
0	0	0	0
0	0	1	0
0	1	0	0
0	1	1	0
1	0	0	0
1	0	1	1
1	1	0	1
1	1	1	0

d.

C	B	A	Y
0	0	0	1
0	0	1	0
0	1	0	1
0	1	1	0
1	0	0	0
1	0	1	0
1	1	0	0
1	1	1	1

4. Sum-of-products
5. Product-of-sums

6. $\bar{C} \cdot B \cdot \bar{A} + C \cdot \bar{B} \cdot A = Y$

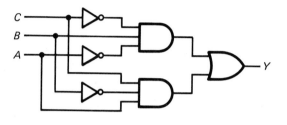

7. *a–c.*

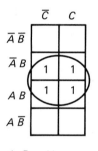

d. $B = Y$

8. *a–c.*

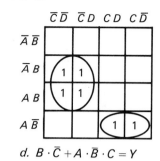

d. $B \cdot \bar{C} + A \cdot \bar{B} \cdot C = Y$

9. *a–c.*

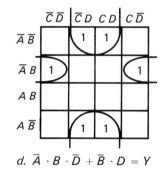

d. $\bar{A} \cdot B \cdot \bar{D} + \bar{B} \cdot D = Y$

10. *a–c.*

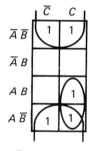

d. $\bar{B} + A \cdot C = Y$

Codes, Encoding, Decoding

- We as humans use the *decimal* code to represent numbers. Digital electronic circuits in calculators or computers use mostly the *binary* code to represent numbers. Both decimal and binary numbers were explored in Chap. 2. Many other special codes are used in digital electronics to represent numbers and even letters of the alphabet. This chapter covers several common *codes* used in digital electronic equipment.

  Electronic translators, which convert from one code to another, are widely used in digital electronics. In Chap. 2 we used an *encoder* to translate from decimal to binary numbers and a *decoder* to translate back from binary to decimal numbers This chapter introduces you to several very common encoders and decoders used for translating from code to code.

5-1 THE 8421 CODE

How would you represent the decimal number 926 in binary form? In other words, how would you convert 926 to the binary number 1110011110? The decimal to binary conversion would be done by using the method from Chap. 2:

Decimal number

926 $\div$ 2 = 463	with a remainder of	0	1s
463 $\div$ 2 = 231	with a remainder of	1	2s
231 $\div$ 2 = 115	with a remainder of	1	4s
115 $\div$ 2 = 57	with a remainder of	1	8s
57 $\div$ 2 = 28	with a remainder of	1	16s
28 $\div$ 2 = 14	with a remainder of	0	32s
14 $\div$ 2 = 7	with a remainder of	0	64s
7 $\div$ 2 = 3	with a remainder of	1	128s
3 $\div$ 2 = 1	with a remainder of	1	256s
1 $\div$ 2 = 0	with a remainder of	1	512s

binary number

The binary number 1110011110 (say "one one one zero zero one one one one zero") does not make much sense to most of us. A code

	HUNDREDS	TENS	ONES
Decimal number	9	2	6
	↓	↓	↓
8421 BCD coded number	1001	0010	0110

Fig. 5-1 Converting from decimal to 8421 code.

that uses binary in a different way is called the 8421 *binary-coded decimal code*. This code is frequently referred to as just the *BCD code*.

The decimal number 926 is converted into the BCD (8421) code in Fig. 5-1. The result is that the decimal number 926 equals 1001 0010 0110 in the 8421 BCD code. Notice from Fig. 5-1 that each group of four binary digits represents a decimal digit. In Fig. 5-1 the right group (0110) represents the 1s place value in the decimal number. The middle group (0010) represents the 10s place value in the decimal number. The left group (1001) represents the 100s place value in the decimal number.

Suppose you were given the 8421 BCD code number 0001 1000 0111 0001. What decimal number does this represent? Figure 5-2 shows how you translate from the BCD code to a decimal number. We find that the BCD number 0001 1000 0111 0001 is equal to the decimal number 1871.

THOUSANDS	HUNDREDS	TENS	ONES
8421 BCD coded number 0001	1000	0111	0001
Decimal number 1	8	7	1

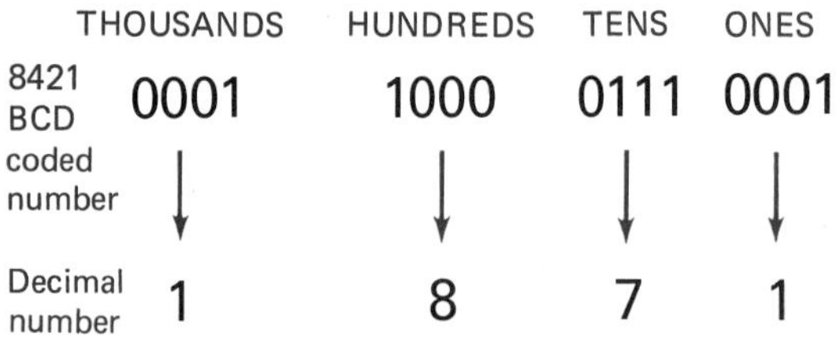

Fig. 5-2 Converting from BCD to decimal numbers.

The 8421 BCD code is very widely used in digital systems. As pointed out, it is common practice to substitute the term BCD code to mean the 8421 BCD code. A word of caution, however: BCD codes do have different weightings of the place values, such as the 4221 code, the excess-3 code, and so on.

5-2 THE EXCESS-3 CODE

The term BCD is a general term, usually referring to an 8421 code. But another code that is really a BCD code is the *excess-3 code*. To convert a decimal number into the excess-3 form we *add 3 to each digit of the decimal number* and convert into binary form. Figure 5-3 shows an example of how the decimal number 4 is converted into the excess-3 code number 0111. Some decimal numbers are converted to excess-3 code in Table 5-1. You

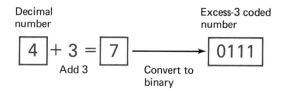

Fig. 5-3 Converting a decimal number to the excess-3 code.

probably have noticed that the code for decimal numbers is rather difficult to figure out. This is because the binary digits are not weighted as they were in regular binary numbers and in the 8421 BCD code. The excess-3 code is used in many arithmetic circuits because it is self-complementing.

The 8421 and the excess-3 codes are but two of many BCD codes used in digital electronics. The 8421 code is by far the most widely used BCD code.

5-3 THE GRAY CODE

Table 5-2 compares the *Gray code* with some codes you already know. The important characteristic of the Gray code is that only *one digit changes* as you *count* from top to bottom, as shown in Table 5-2. The Gray code cannot be used in arithmetic circuits. The Gray code is used for input and output devices

Table 5-1 The Excess-3 code.

Decimal number	Excess-3 coded number		
0			0011
1			0100
2			0101
3			0110
4			0111
5			1000
6			1001
7			1010
8			1011
9			1100
14		0100	0111
27		0101	1010
38		0110	1011
459	0111	1000	1100
606	1001	0011	1001
	Hundreds	Tens	Ones

Table 5-2 The Gray code.

Decimal number	Binary number	BCD 8421 coded number		Gray coded number
0	0000		0000	0000
1	0001		0001	0001
2	0010		0010	0011
3	0011		0011	0010
4	0100		0100	0110
5	0101		0101	0111
6	0110		0110	0101
7	0111		0111	0100
8	1000		1000	1100
9	1001		1001	1101
10	1010	0001	0000	1111
11	1011	0001	0001	1110
12	1100	0001	0010	1010
13	1101	0001	0011	1011
14	1110	0001	0100	1001
15	1111	0001	0101	1000
16	10000	0001	0110	11000
17	10001	0001	0111	11001

in digital systems. You can see from Table 5-2 that the Gray code is not classed as one of the many BCD codes. And notice that it is quite difficult to translate from decimal numbers to the Gray code and back to decimals again. There is a method for making this conversion, but we usually use electronic decoders to do the job for us.

5-4 ENCODERS

A digital system using an *encoder* is shown in Fig. 5-4. The encoder in this system must translate the decimal input from the keyboard to an 8421 BCD code. You used an encoder of this type in the experiments in Chap. 2. This encoder is called a *10-line-to-4-line priority encoder* by the manufacturer. Figure 5-5 is a block diagram of this encoder. If the decimal input 3 on the encoder is activated, then the logic circuits inside the unit outputs the BCD 0011 as shown.

The encoder is a logic circuit with nine inputs and four outputs. This encoder uses about 30 logic gates to perform its job. For convenience, manufacturers package this encoder in a single IC.

5-5 DECODERS

A *decoder*, like an encoder, is a code translator. Figure 5-4 shows that two decoders are being used in this system. The decoders are translating the 8421 BCD code to a seven-segment display code that lights the proper segments on the displays. The display will be

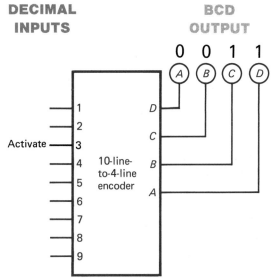

DECIMAL INPUTS

BCD OUTPUT

Fig. 5-5 10-line-to-4-line encoder.

10-line-to-4-line priority encoder

BCD-to-seven-segment decoder

Seven-segment LED display

a decimal number. Figure 5-6 shows the BCD 0101 at the input of the *BCD-to-seven-segment decoder*. The decoder activates outputs *a*, *c*, *d*, *f*, and *g* to light the segments shown in Fig. 5-6. The decimal number 5 lights up on the display.

Each segment of the seven-segment display is lettered with lowercase letters *a* to *g*, as shown in Fig. 5-6. The displays you will use are LED (light-emitting diode) displays. Each segment is an LED which gives off a reddish glow when lit. Figure 5-7 illustrates how the numbers 0 to 15 are displayed on a seven-segment LED display. Normally the 10 to 15 displays are not used. The seg-

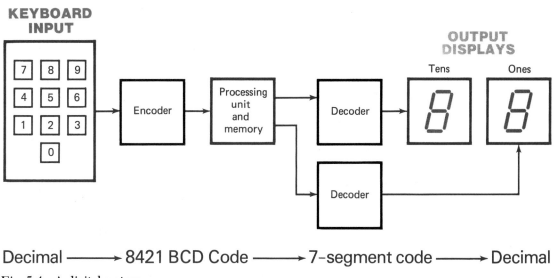

Fig. 5-4 A digital system.

43

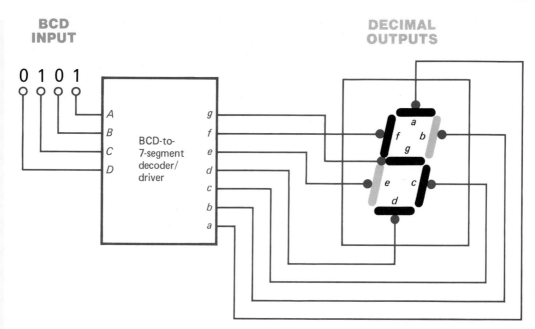

Fig. 5-6   Decoder driving a seven-segment display.

Fig. 5-7   Number readouts on seven-segment display.

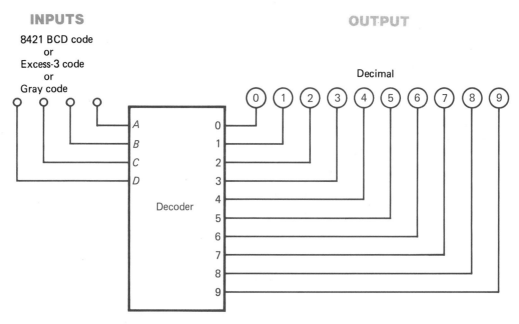

Fig. 5-8   A typical decoder block diagram. Note that the inputs may be in either the 8421, excess-3, or Gray code.

44

ments of the seven-segment may be lit by using other methods. Displays come in fluorescent, incandescent, LCD (liquid-crystal display), and gas-discharge types.

Decoders come in several varieties, such as the ones illustrated in Fig. 5-8. Notice in Fig. 5-8 that the same block diagram is used for the 8421 BCD, the excess-3, and the Gray decoders.

Decoders, like encoders, are combinational logic circuits with several inputs and outputs. Most decoders contain from 20 to 50 gates. Most decoders and encoders are packaged in single IC packages.

Summary

1. Many codes are used in digital equipment. You will be familiar with decimal, binary, 8421 BCD, excess-3, and Gray codes.
2. Converting from code to code is essential for your work in digital electronics. Table 5-3 will aid you in converting from one code to another.
3. Electronic translators are called encoders and decoders. These complicated logic circuits are manufactured in single IC packages.
4. Seven-segment displays are very popular devices for reading out numbers. There are decoders that will directly drive most seven-segment displays.

Table 5-3   Summary of common codes.

Decimal number	Binary number	BCD codes 8421		BCD codes Excess-3		Gray code
0	0000		0000		0011	0000
1	0001		0001		0100	0001
2	0010		0010		0101	0011
3	0011		0011		0110	0010
4	0100		0100		0111	0110
5	0101		0101		1000	0111
6	0110		0110		1001	0101
7	0111		0111		1010	0100
8	1000		1000		1011	1100
9	1001		1001		1100	1101
10	1010	0001	0000	0100	0011	1111
11	1011	0001	0001	0100	0100	1110
12	1100	0001	0010	0100	0101	1010
13	1101	0001	0011	0100	0110	1011
14	1110	0001	0100	0100	0111	1001
15	1111	0001	0101	0100	1000	1000
16	10000	0001	0110	0100	1001	11000
17	10001	0001	0111	0100	1010	11001
18	10010	0001	1000	0100	1011	11011
19	10011	0001	1001	0100	1100	11010
20	10100	0010	0000	0101	0011	11110

Questions

1. Write the binary number for the decimal numbers in *a* to *f*:
   a. 17   *c.* 42   *e.* 150
   b. 31   *d.* 75   *f.* 300

2. Write the 8421 BCD coded number for the decimal numbers in *a* to *f*:
   a. 17   *d.* 1632
   b. 31   *e.* 47,899
   c. 150   *f.* 103,926

3. Write the decimal number for the 8421 BCD coded numbers in *a* to *h*:
   a. 0010
   b. 1111
   c. 0011 0000
   d. 1110 0000 1111
   e. 0111 0001 0110 0000
   f. 0001 0001 0000 0000 0000
   g. 0101 1001 1000 1000 0101
   h. 0011 0010 0001 0100 0101 0110

4. Write the binary number for the 8421 BCD coded numbers in question 3.

5. Write the excess-3 coded number for the decimal numbers in *a* to *f*:
   a. 7   *d.* 318
   b. 27   *e.* 4063
   c. 59   *f.* 5533

6. Why is the excess-3 code used in some arithmetic circuits?

7. List two codes you learned about that are classed as BCD codes.

8. Write the Gray code number for the decimal numbers in *a* to *f*:
   a. 1   *d.* 4
   b. 2   *e.* 5
   c. 3   *f.* 6

9. As you count in the Gray code, what is the most important characteristic of this code?

10. List two general names for code translators or electronic code converters.

11. A(n) ____?____ (decoder, encoder) is the electronic device that would be used to convert the decimal input of a calculator to the BCD code used by the central processing unit.

12. A(n) ____?____ (decoder, encoder) is the electronic device that would be used to convert the BCD of the central processing unit of a calculator to the decimal-display output.

13. Draw a seven-segment display and label each segment (use the letters a, b, c, d, e, f, and g).

14. Which segments of the seven-segment display *will light* when the following decimal numbers appear? Use the letters a, b, c, d, e, f, and g as answers.
    a. 0   *f.* 5
    b. 1   *g.* 6
    c. 2   *h.* 7
    d. 3   *i.* 8
    e. 4   *j.* 9

15. The seven-segment displays that you will use give off a red glow and are of what type?

Flip-Flops

- Engineers classify logic circuits into two groups. We already worked with *combinational logic* circuits using AND, OR, and NOT gates. The other group of circuits are classified as *sequential logic* circuits. Sequential circuits involve timing and memory devices.

  The basic building block for combinational logic circuits is the logic gate. The basic building block for sequential logic circuits is the *flip-flop*. This chapter covers several types of flip-flop circuits. In later chapters you will wire flip-flops together. Flip-flops are wired to form counters, shift registers, and various memory devices.

6-1 THE R-S FLIP-FLOP

The logic symbol for the *R-S flip-flop* is drawn in Fig. 6-1. Notice that the R-S flip-flop has two inputs labeled S and R. The two outputs are labeled Q and $\bar{Q}$ (say "not Q"). In flip-flops the outputs are always opposite or complementary. In other words, if output $Q = 1$, then output $\bar{Q} = 0$, and so on. The letters S and R at the inputs of the R-S flip-flop are often referred to as the *set* and *reset* inputs.

The truth table in Table 6-1 shows the detailed operation of the R-S flip-flop. When the S and R inputs are both 0, both outputs go to a logical 1. This is called a *prohibited state* for the flip-flop and is not used. The second line shows that when input S is 0 and R is 1, the output Q will be set to logical 1. The third line shows that when input R is 0 and S is 1, output Q will be reset (or cleared) to 0. Line four shows both inputs (R, S) at 1. This is

Table 6-1   Truth table for R-S flip-flop.

INPUTS		OUTPUTS		
S	**R**	**Q**	**$\bar{Q}$**	**Effect on output Q**
0	0	1	1	Prohibited - Do not use
0	1	1	0	For setting Q to 1
1	0	0	1	For resetting Q to 0
1	1	Q	$\bar{Q}$	Depends on previous state

the idle or at rest condition and leaves Q and $\bar{Q}$ in their previous complementary states.

If you observe very carefully, you will see that it takes a logical 0 to activate the set (set Q to 1). It also takes a logical 0 to activate the reset or clear (clear Q to 0). Because it takes a logical 0 to enable or activate the flip-flop, the logic symbol in Fig. 6-2(*a*) is probably more accurate. Notice the invert bubbles at the R and S inputs. These invert bubbles indicate that the set and reset inputs are activated by a logical 0.

R-S flip-flops can be purchased in an IC package, or they can be wired from logic gates, as shown in Fig. 6-2(*b*). The NAND gates in Fig. 6-2(*b*) form an R-S flip-flop. This NAND gate R-S flip-flop operates according to the truth table in Table 6-1.

Many times *timing diagrams* or *waveforms* are given for sequential logic circuits. These diagrams show the voltage level and timing between inputs and outputs and are similar to

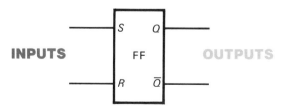

Fig. 6-1   Logic symbol for an R-S flip-flop.

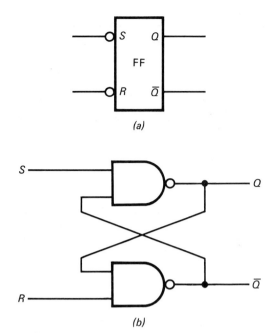

Fig. 6-2 (*a*) **R-S flip-flop.** (*b*) **Wiring an R-S flip-flop using NAND gates.**

what you would observe on an oscilloscope. The horizontal distance is *time*, and the vertical distance is *voltage*. Figure 6-3 shows the input waveforms (*R*, *S*) and the output waveforms (*Q*, $\bar{Q}$) for the R-S flip-flop. The bottom of the diagram lists the lines of the truth table from Table 6-1. The *Q* waveform shows the set and reset conditions of the output; the logic levels (0, 1) are on the right side of the

waveforms. Waveform diagrams of the type shown in Fig. 6-3 are very common when dealing with sequential logic circuits. Study this diagram to see what it tells you. The waveform diagram is really a type of truth table.

The R-S flip-flop is also called an *R-S latch* or *set-reset flip-flop*. Do you know the logic symbol and truth table for the R-S latch or R-S flip-flop?

6-2 THE CLOCKED R-S FLIP-FLOP

The logic symbol for a *clocked R-S flip-flop* is shown in Fig. 6-4. Observe that it looks just like an R-S flip-flop except that it has one extra input labeled *CLK* (for clock). Figure 6-5 diagrams the operation of the clocked R-S flip-flop. The *CLK* input is at the top of the diagram. Notice that the clock pulse (1) has no effect on output *Q* with inputs *S* and *R* in the 0 position. At the preset *S* position the *S* (set) input is moved to 1, but output *Q* will not be set to 1 until clock pulse 2 permits *Q* to go to 1. Once the flip-flop latch has been set, more pulses (pulses 3 and 4) will not affect output *Q*. Next, input *R* is preset to 1. At the next clock pulse (5) the *Q* output is reset (or cleared) to 0. Output *Q* stays reset (at 0) even if more pulses (pulses 6 and 7) appear at the clock input.

Notice that the outputs of the clocked R-S

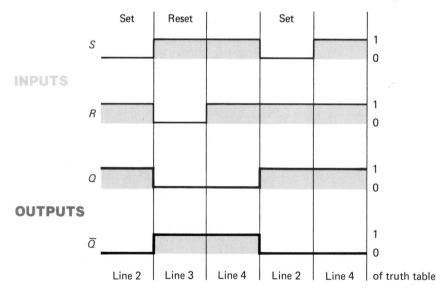

Fig. 6-3 Waveform diagram for an R-S flip-flop.

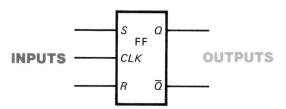

INPUTS ──── OUTPUTS

Fig. 6-4   Logic symbol for a clocked R-S flip-flop.

flip-flop *change only on a clock pulse*. We say that this flip-flop operates *synchronously*; it operates in *step with* the clock. Synchronous operation is very important in calculators and computers, where each step must happen in a very exact order.

Another characteristic of the clocked R-S flip-flop is that once it is set or reset it stays that way even if you change some inputs. This is a *memory characteristic*, which is extremely valuable in many digital circuits.

Figure 6-6(*a*) shows a truth table for the clocked R-S flip-flop. Notice that only the top three lines of the truth table are usable; the bottom line is prohibited and not used.

Figure 6-6(*b*) shows a wiring diagram of a clocked R-S flip-flop. Notice that two NAND gates were added to the inputs of the R-S flip-flop to add the clocked feature.

It is highly suggested that you actually wire the R-S and clocked R-S flip-flops. Operating flip-flops in the lab will help you to better understand their operation.

TRUTH TABLE

INPUTS			OUTPUTS		
CLK	**S**	**R**	**Q**	**Q̄**	**Effect on output Q**
⊓	0	0	No change	No change	
⊓	0	1	0	1	Reset or cleared to 0
⊓	1	0	1	0	Set to 1
⊓	1	1	1	1	Prohibited— do not use

(a)

Synchronous

Memory characteristic

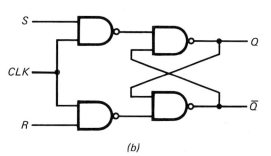

(b)

Fig. 6-6   (*a*) Truth table for a clocked R-S flip-flop.   (*b*) Wiring a clocked R-S flip-flop.

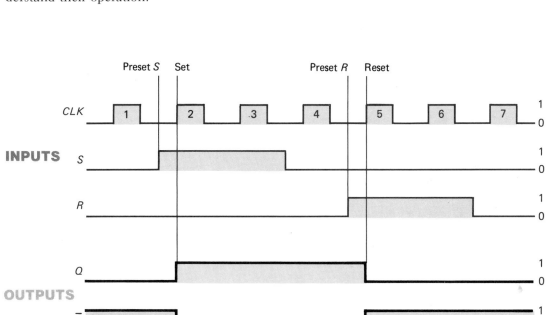

Fig. 6-5   Waveform diagram for a clocked R-S flip-flop

49

6-3 THE D FLIP-FLOP

The logic symbol for the D *flip-flop* is shown in Fig. 6-7(a). It has only one *data input* (D) and a clock input (CLK). The outputs are labeled Q and $\bar{Q}$. The D flip-flop is often called a *delay flip-flop*. The word "delay" describes what happens to the data or information at input D. The data (a 0 or 1) at input D is *delayed one clock pulse* from getting to output Q. A simplified truth table for the D flip-flop is shown in Fig. 6-7(b). Notice that output Q follows input D *after one clock pulse* (see Q^{n+1} column).

A D flip-flop may be formed from a clocked R-S flip-flop by adding an inverter, as shown in Fig. 6-8(a). More commonly you will use a D flip-flop contained in an IC. Figure 6-8(b) shows a typical commercial D flip-flop. Two extra inputs [PS (preset) and CLR (clear)] have been added to the D flip-flop in Fig. 6-8(b). The PS input will set output Q to 1 when enabled by a logical 0. The CLR input will clear output Q to a 0 when enabled by a logical 0. The inputs PS and CLR will override the D and CLK inputs. The D and CLK inputs operate as they did in the D flip-flops in Fig. 6-7.

D flip-flops are wired together to form *shift registers* and *storage registers*. These registers

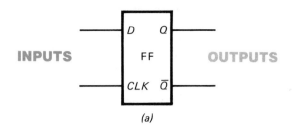

(a)

Truth table

Input	Output
D	Q^{n+1}
0	0
1	1

(b)

Fig. 6-7 D Flip-flop. (a) **Logic symbol.** (b) **Truth table.**

are widely used in digital systems. Remember the D flip-flop *delays* data from reaching output Q one clock pulse and is called a delay flip-flop.

6-4 THE J-K FLIP-FLOP

The *J-K flip-flop* is probably the most widely used and universal flip-flop, having the features of all the other types of flip-flops. The

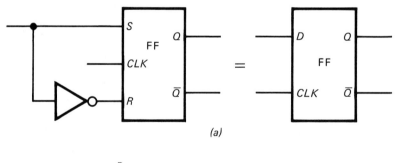

(a)

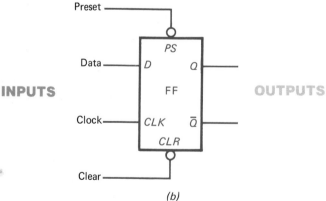

(b)

Fig. 6-8 (a) **Wiring a D flip-flop.** (b) **Commercial D-type flip-flop.**

D flip-flop

Delay flip-flop

J-K flip-flop

Shift register

Storage register

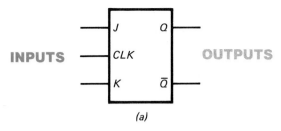

(a)

Truth table

INPUTS			OUTPUTS		
CLK	J	K	Q	$\bar{Q}$	Effect on output Q
⎍	0	0	No change		No change — disable
⎍	0	1	0	1	Reset or cleared to 0
⎍	1	0	1	0	Set to 1
⎍	1	1	Toggle		Changes to opposite state

(b)

Fig. 6-9 J-K flip-flop. (a) **Logic symbol.** (b) **Truth table.**

logic symbol for the J-K flip-flop is illustrated in Fig. 6-9(a). The inputs labeled J and K are the data inputs. The input labeled CLK is the clock input. Outputs Q and $\bar{Q}$ are the usual complementary outputs on a flip-flop. A truth table for the J-K flip-flop is shown in Fig. 6-9(b). When the J and K inputs are both 0, the flip-flop is *disabled* and the outputs do not change states. Lines 2 and 3 of the truth table show the reset and set conditions for the Q output. Line 4 illustrates the useful *toggle* position of the J-K flip-flop. When both data

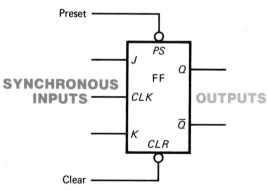

Fig. 6-10 Commercial J-K flip-flop with PS and CLR inputs.

inputs J and K are at 1, repeated clock pulses cause the output to turn off-on-off-on-off-on, and so on. This off-on-off-on action is like a toggle switch and so is called *toggling*.

Figure 6-10 is the logic symbol for a commercial J-K flip-flop. The synchronous inputs J, K, and CLK work according to the truth table in Fig. 6-9(b). The other *asynchronous* (not synchronous—not in step with the clock) inputs are for setting or clearing the Q output of the flip-flop. The PS and CLR inputs override the synchronous inputs. The asynchronous inputs PS and CLR work exactly like the R and S inputs on an R-S flip-flop (see Table 6-1).

J-K flip-flops are widely used in many digital circuits. You will use the J-K flip-flop especially in *counters*. Counters are found in almost every digital system.

6-5 TRIGGERING FLIP-FLOPS

We have classified flip-flops as synchronous or asynchronous in their operation. Synchronous flip-flops are all those that have a clock input. We found that the clocked R-S, the D, and the J-K flip-flops operate in step with the clock.

When using manufacturers' data manuals you will notice that many synchronous flip-flops are also classified as either *edge-triggered* or *master/slave*. Figure 6-11 shows two edge-triggered flip-flops in the toggle position. On clock pulse 1 the positive edge (positive-going edge) of the pulse is identified. The second waveform shows how the positive-edge-triggered flip-flop toggles each time a positive-going pulse comes along (see pulses 1 to 4). On pulse 1 on Fig. 6-11 the negative-edge (negative-going edge) of the pulse is also labeled. The bottom waveform shows how the negative-edge-triggered flip-flop would toggle. Notice that it changes state or toggles each time a negative-going pulse comes along (see pulses 1 to 4). And especially notice the difference in timing between the positive- and negative-edge-triggered flip-flops. This triggering time difference is quite important for some applications.

Another class of flip-flop triggering is the master/slave type. The J-K master/slave flip-flop uses the entire pulse (positive-edge and negative-edge) to trigger the flip-flop. Figure 6-12 shows the triggering of a master/slave flip-

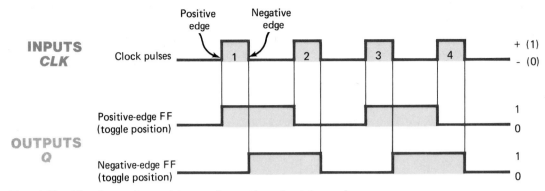

Fig. 6-11 Waveforms for positive- and negative-edge-triggered flip-flops.

flop. Pulse 1 shows four positions (*a* to *d*) on the waveform. The following sequence of operation takes place in the master/slave flip-flop at each point on the clock pulse:

- Point *a*: leading edge—isolate input from output
- Point *b*: leading edge—enter information from *J* and *K* inputs
- Point *c*: trailing edge—disable *J* and *K* inputs
- Point *d*: trailing edge—transfer information from input to output

A very interesting characteristic of the master/slave flip-flop is shown on pulse 2, Fig. 6-12. Notice that at the beginning of pulse 2 the outputs are disabled. For a very brief moment the *J* and *K* inputs are moved to the toggle position (see point *e*) and then disabled. The J-K master/slave flip-flop "remembers" that the *J* and *K* inputs were in the toggle position, and it toggles at point *f* on the waveform diagram. This memory characteristic will only happen while the clock pulse is high (at logical 1).

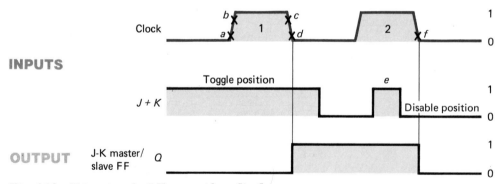

Fig. 6-12 Triggering the J-K master/slave flip-flop.

Summary

1. Logic circuits are classified as combinational or sequential. Combination logic circuits use AND, OR, and NOT gates. Sequential logic circuits use flip-flops and involve a memory characteristic.
2. Flip-flops are wired together to form counters, registers, and memory devices.

3. Flip-flop outputs are always opposite or complementary.
4. Table 6-2 summarizes some basic flip-flops.
5. Waveform (timing) diagrams are used to describe the operation of sequential devices.
6. Flip-flops can be edge-triggered or master/slave types.

Table 6-2 Summary of basic flip-flops.

Circuit	Logic Symbol	Truth table	Remarks:
R-S flip-flop	S Q / FF / R Q̄	S R Q 0 0 prohibited 0 1 1 set 1 0 0 reset 1 1 disable	R-S latch Set-reset flip-flop (asynchronous)
Clocked R-S flip-flop	S Q / FF / CLK / R Q̄	CLK S R Q ⊓ 0 0 disable ⊓ 0 1 0 reset ⊓ 1 0 1 set ⊓ 1 1 prohibited	(synchronous)
D flip-flop	D Q / FF / CLK Q̄	CLK D Q ⊓ 0 0 ⊓ 1 1	Delay flip-flop Data flip-flop (synchronous)
J-K flip-flop	J Q / FF / CLK / K Q̄	CLK J K Q ⊓ 0 0 disable ⊓ 0 1 0 ⊓ 1 0 1 ⊓ 1 1 toggle	Most universal FF (synchronous)

Questions

1. Logic ___?___ are the basic building blocks of combinational logic circuits; the basic building blocks of sequential circuits are devices called ___?___.

2. List one type of asynchronous and three types of synchronous flip-flops (mark the synchronous types).

3. List two other names sometimes given to a R-S flip-flop.

4. Draw a logic symbol for the following flip-flops:
   a. J-K   c. Clocked R-S
   b. D   d. R-S

5. Draw a truth table for the following flip-flops:
   a. J-K   c. Clocked R-S
   b. D   d. R-S

6. Which flip-flop has no prohibited states in its truth table?

7. The ___?___ flip-flop is the most widely used.

8. Draw the logic symbol for the following flip-flops that have *PS* and *CLR* asynchronous inputs:
   a. D *b.* J-K

9. Draw the truth table for the asynchronous inputs only (*PS* and *CLR*) on the following flip-flops:
   a. D *b.* J-K

10. If both the synchronous and asynchronous inputs on a J-K flip-flop are activated, which input will control the output?

11. When we say the flip-flop is in the set condition, we mean output ___?___ is at a logical ___?___.

12. When we say the flip-flop is in the reset or clear condition, we mean output ___?___ is at a logical ___?___.

13. On a timing or waveform diagram the horizontal distance stands for ___?___, and the vertical distance stands for ___?___.

14. Refer to Fig. 6-3. Notice that line 4 is listed two times across the bottom. Why does output $Q = 0$ once and $Q = 1$ on the right side when inputs *R* and *S* are both 1 in each case?

15. Refer to Fig. 6-5. This waveform diagram is for a ___?___ flip-flop. This flip-flop is ___?___-edge-triggered.

16. List two types of edge-triggered flip-flops.

17. If a flip-flop is in its toggle position, what will the output act like upon repeated clock pulses?

18. Explain how a master/slave J-K flip-flop is triggered.

19. Identify these acronyms used on flip-flops:
   a. CLK   *e.* PS
   b. CLR   *f.* R
   c. D   *g.* S
   d. FF

20. Explain how a master/slave J-K flip-flop can still toggle even when *J* and *K* inputs are in the "disable" condition?

Counters

- Almost any complex digital system contains several *counters*. A counter's job is the obvious one of counting events or periods of time or putting events into sequence. Counters also do some not-so-obvious jobs: dividing frequency, addressing, and serving as memory units. This chapter discusses several types of counters and their uses.
    Flip-flops are wired together to form circuits that count. Because of the wide use of counters, manufacturers also make self-contained counters in IC form.

7-1 THE RIPPLE COUNTER

Counting in binary and decimal is illustrated in Fig. 7-1. With the four binary places (D, C, B, and A) we can count from 0000 to 1111 (0 to 15 in decimal). Notice that column A is the 1s binary place or least significant digit (LSD). The term least significant bit (LSB) is also used. Column D is the 8s binary place or the most significant digit (MSD). The term most significant bit (MSB) is also used. Notice that the 1s column changes state the most often. If we design a counter to count from binary 0000 to 1111, we need a device that will have 16 different output states: a *modulo (mod)-16 counter*. The modulus of a counter is the number of different states the counter must go through to complete its counting cycle.

A modulo-16 counter using four J-K flip-flops is diagramed in Fig. 7-2(a). Each J-K flip-flop is in its toggle position (J and K both at 1). Assume the outputs are cleared to 0000. As clock pulse 1 arrives at the clock (CLK) input of FF 1 it will toggle (on the negative edge), and the display will show 0001. Clock pulse 2 will cause FF 1 to toggle again, returning output Q to 0, which will cause FF 2 to toggle to 1. The count on the display will now read 0010. The counting will continue, with each flip-flop output triggering the next flip-flop on its negative-going pulse. Look back at Fig. 7-1 and see that column A (1s col-umn) must change state on every count. This means that FF 1 in Fig. 7-2(a) must toggle each pulse. FF 2 must toggle only half as often as FF 1, as seen from column B in Fig. 7-1. Each more significant digit in Fig. 7-1 toggles less often.

The counting of the modulo-16 counter is

BINARY COUNTING				DECIMAL COUNTING
D	C	B	A	
8s	4s	2s	1s	
0	0	0	0	0
0	0	0	1	1
0	0	1	0	2
0	0	1	1	3
0	1	0	0	4
0	1	0	1	5
0	1	1	0	6
0	1	1	1	7
1	0	0	0	8
1	0	0	1	9
1	0	1	0	10
1	0	1	1	11
1	1	0	0	12
1	1	0	1	13
1	1	1	0	14
1	1	1	1	15

Fig. 7-1 Counting sequence of a counter.

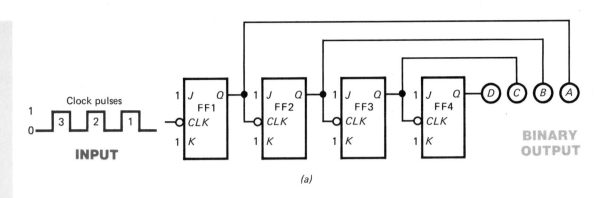

INPUT

(a)

From page 55:
Modulus of a
counter

On this page:
Ripple counter

Modulo-10
counter

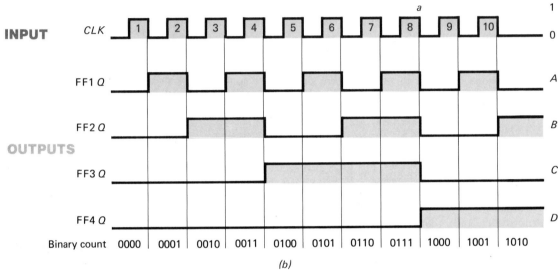

(b)

Fig. 7-2 Modulo-16 counter. (*a*) Logic diagram. (*b*) Waveform
diagram.

shown up to a count of 10 (1010) by wave-forms in Fig. 7-2(*b*). The *CLK* input is shown on the top line. The state of each flip-flop (FF 1, FF 2, FF 3, FF 4) is shown on the wave-forms below. The binary count is shown across the bottom of the diagram. Especially note the vertical lines on Fig. 7-2(*b*); these lines show that the clock triggers only FF 1. FF 1 triggers FF 2, FF 2 triggers FF 3, and so on. Because one flip-flop affects the next one, it does take some time to toggle all the flip-flops. For instance, at point (*a*) on pulse 8, Fig. 7-2(*b*), notice that the clock triggers FF 1, causing it to go to 0. This in turn causes FF 2 to toggle from 1 to 0. This in turn causes FF 3 to toggle from 1 to 0. As output *Q* of FF 3 reaches 0 it triggers FF 4, which toggles from 0 to 1. We see that the changing of states is a chain reaction that *rip-*

ples through the counter. For this reason this counter is called a *ripple counter*.

The counter we studied in Fig. 7-2 could be described as a ripple counter, a modulo-16 counter, a 4-bit binary counter, or an asynchronous counter. All these names de-scribe something about the counter. The rip-ple and asynchronous labels mean that all the flip-flops do not trigger at one time. The modulo-16 description comes from the num-ber of states the counter goes through. The 4-bit label tells how many binary places there are at the output of the counter.

7-2 MODULO-10 RIPPLE COUNTER

The counting sequence for a modulo-10 counter is from 0000 to 1001 (0 to 9 in deci-

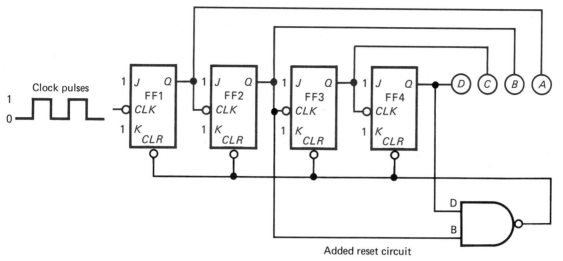

Fig. 7-3   Logic diagram of a modulo-10 ripple counter.

mal).   This is down to the heavy line in Fig. 7-1.   This mod-10 counter, then, has four place values: 8s, 4s, 2s, 1s.   This takes four flip-flops connected as a ripple counter in Fig. 7-3).   We must *add* a NAND gate to the ripple counter to clear all the flip-flops back to zero *immediately after* the 1001 (9) count. The trick is to look at Fig. 7-1 and determine what the next count will be after 1001.   You will find it is 1010 (10).   You must feed the two 1s in the 1010 into a NAND gate as shown in Fig. 7-3.   The NAND gate will then clear the flip-flop back to 0000.   The counter will then start its count from 0000 up to 1001 again. We say we are using the NAND gate to reset the counter to 0000.   By using a NAND gate in this manner we can make several other modulo counters.   Figure 7-3 illustrates a mod-10 ripple counter.   This type of counter might also be called a *decade* (meaning 10) *counter*.

Ripple counters can be constructed from individual flip-flops.   Manufacturers also produce ICs with all four flip-flops inside a single package.   Some IC counters even contain the reset NAND gate, such as the one you used in Fig. 7-3.

7-3 SYNCHRONOUS COUNTERS

The ripple counters we studied were asynchronous counters.   Each flip-flop did not trigger exactly in step with the clock pulse.

For some high frequency operations it is necessary to have all stages of the counter trigger together.   There is such a counter: a *synchronous counter*.

A rather complicated-looking synchronous counter is shown in Fig. 7-4(*a*).   This logic diagram is for a 3-bit (mod-8) counter.   First notice the *CLK* connections.   The clock is connected directly to the *CLK* input of each flip-flop.   We say that the *CLK* inputs are connected in *parallel*.   Figure 7-4(*b*) gives the counting sequence this counter will go through.   Column *A* is the binary 1s column, and FF 1 does the counting for this column. Column *B* is the binary 2s column, and FF 2 counts this column.   Column *C* is the binary 4s column, and FF 3 counts this column.

Let us go through the counting sequence of this mod-8 counter by referring to Fig. 7-4(*a*) and (*b*):

Pulse 1–row 2
   Circuit action: Each FF is pulsed by clock. Only FF 1 can toggle because it is the only one with 1s applied to both *J* and *K* inputs.
   Output result: 001 (decimal 1)

Pulse 2–row 3
   Circuit action: Each FF is pulsed.
      Two FFs will toggle because they have 1s applied to both *J* and *K* inputs.
      FF 1 and FF 2 will both toggle.
      FF 1 goes from 1 to 0.
      FF 2 goes from 0 to 1.
   Output result: 010 (decimal 2)

57

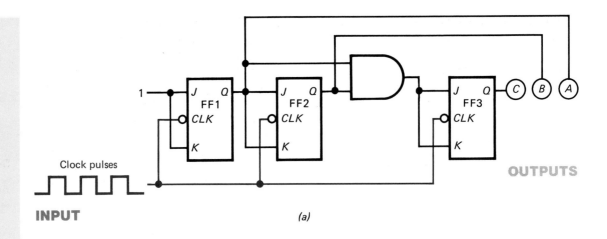

Clock pulses

INPUT

OUTPUTS

(a)

ROW	NUMBER OF CLOCK PULSES	BINARY COUNTING SEQUENCE			DECIMAL COUNT
		C	B	A	
1	0	0	0	0	0
2	1	0	0	1	1
3	2	0	1	0	2
4	3	0	1	1	3
5	4	1	0	0	4
6	5	1	0	1	5
7	6	1	1	0	6
8	7	1	1	1	7
9	8	0	0	0	0

(b)

Fig. 7-4 3-bit synchronous counter. (a) Logic diagram.
(b) Counting sequence.

Pulse 3–row 4
 Circuit action: Each FF is pulsed.
  Only one FF toggles.
  FF 1 toggles from 0 to 1.
 Output result: 011 (decimal 3)

Pulse 4–row 5
 Circuit action: Each FF is pulsed.
  All FFs toggle to opposite state.
  FF 1 goes from 1 to 0.
  FF 2 goes from 1 to 0.
  FF 3 goes from 0 to 1.
 Output result: 100 (decimal 4)

Pulse 5–row 6
 Circuit action: Each FF is pulsed.
  Only one FF toggles.
  FF 1 goes from 0 to 1.
 Output result: 101 (decimal 5)

Pulse 6–row 7
 ·*Circuit action:* Each FF is pulsed.
  Two FFs toggle.
  FF 1 goes from 1 to 0.
  FF 2 goes from 0 to 1.
 Output result: 110 (decimal 6)

Pulse 7–row 8
 Circuit action: Each FF is pulsed.
  FF 1 goes from 0 to 1.
 Output result: 111 (decimal 7)

Pulse 8–row 9
 Circuit action: Each FF is pulsed.
  All three FFs toggle.
  All FFs change from 1 to 0.
 Output result: 000 (decimal 0)

You now have completed the explanation of
how the 3-bit synchronous counter works.

Notice that the J-K flip-flops were used in their toggle position (J and K at 1) or disable position (J and K at 0) only.

Because of their complexity, synchronous counters are most often purchased in IC form.

7-4 DOWN COUNTERS

To now we have used counters that count upward (0, 1, 2, 3, 4, . . .). But sometimes we must count downward (9, 8, 7, 6, . . .) in digital systems. A counter that counts from higher to lower numbers is called a *down counter*.

A logic diagram of a mod-8 asynchronous down counter is shown in Fig. 7-5(*a*); the counting sequence for this counter is listed in Fig. 7-5(*b*). Note how much the down counter in Fig. 7-5(*a*) looks like the up counter in Fig.

7-2(*a*). The only difference is in the "carry" from FF 1 to FF 2 and the carry from FF 2 to FF 3. The up counter carries from Q to the CLK input of the next flip-flop. The down counter carries from $\bar{Q}$ (not Q) to the CLK input of the next flip-flop. Notice that the down counter has a preset (*PS*) control to preset the counter to 111 (decimal 7) to start the downward count. FF 1 is the binary 1s place (column A) counter. FF 2 is the 2s place (column B) counter. FF 3 is the 4s place (column C) counter.

7-5 SELF-STOPPING COUNTERS

The down counter you studied in Fig. 7-5(*a*) will continue to *recirculate*. That is, when it gets to 000 it will start at 111, then 110, and so forth. However, sometimes you want a

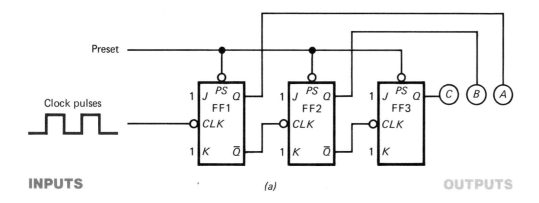

INPUTS (a) OUTPUTS

NUMBER OF CLOCK PULSES	BINARY COUNTING SEQUENCE			DECIMAL COUNT
	C	B	A	
0	1	1	1	7
1	1	1	0	6
2	1	0	1	5
3	1	0	0	4
4	0	1	1	3
5	0	1	0	2
6	0	0	1	1
7	0	0	0	0
8	1	1	1	7
9	1	1	0	6

(b)

Fig. 7-5   3-bit ripple down counter.   (*a*) Logic diagram.
(*b*) Counting sequence.

59

Frequency
division

Divide-by-10
counter

7493 4-bit binary
counter

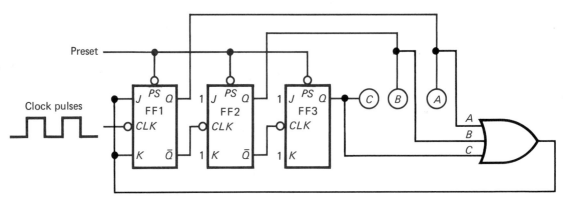

Fig. 7-6   3-bit down counter with self-stopping feature.

counter to *stop* when a sequence is finished. Figure 7-6 illustrates how you could stop the down counter in Fig. 7-5 at the 000 count. The counting sequence is shown in Fig. 7-5(*b*). In Fig. 7-6 we added an OR gate to place a logical 0 on the *J* and *K* inputs of FF 1 when the count at outputs *C*, *B*, and *A* reaches 000. The preset would have to be enabled (*PS* to 0) again to start the sequence at 111 (decimal 7).

Up or down counters can be stopped after any sequence of counts by using a logic gate or combination of gates. The output of the gate is fed back to the *J* and *K* inputs of the first flip-flop in a ripple-type counter.

7-6 COUNTERS AS FREQUENCY DIVIDERS

An interesting and common use of counters is for *frequency division*. An example of a simple system using a frequency divider is shown in Fig. 7-7. This system is the basis for an electric clock. The 60-hertz (Hz) input frequency is from the power line (formed into a square wave). The circuit must divide the frequency by 60, and the output will be one pulse per second (1 Hz). This would be a seconds timer.

A block diagram of a decade counter is drawn in Fig. 7-8(*a*). In Fig. 7-8(*b*) the waveforms at the *CLK* input and the binary 8s place (output Q_D) are shown. Notice that it takes 30 input pulses to produce three output pulses. Using division, we find that $30 \div 3 = 10$. Output Q_D of the decade counter in Fig. 7-8(*a*) would be a *divide-by-10* counter. In other words, the output frequency at Q_D is only one-tenth the frequency at the input of the counter.

If we used the decade counter (divide-by-10 counter) from Fig. 7-8 and a mod-6 counter (divide-by-6 counter) in series, we would get the divide-by-60 circuit we needed in Fig. 7-7. A diagram of such a system is illustrated in Fig. 7-9. The 60-Hz square wave enters the divide-by-6 counter (mod-6 counter) and comes out at 10 Hz. The 10 Hz then enters the divide-by-10 counter (decade counter) and comes out at 1 Hz.

You are already aware that counters are used as frequency dividers in digital timepieces such as electronic digital clocks, automobile digital clocks, and digital wristwatches. Frequency division is also used in frequency counters, oscilloscopes, and television servicing dot-and-bar generators.

7-7 IC COUNTERS

The contents of manufacturers' IC data manuals contain long lists of counters. This unit covers only two representative types of IC counters.

Signetics, a manufacturer of integrated circuits, provided the diagrams and tables in Fig. 7-10. The diagrams are for a 7493 *4-bit binary counter.* Look carefully at the logic diagram

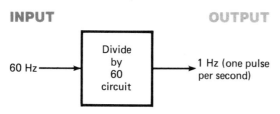

INPUT   **OUTPUT**

60 Hz ⟶ | Divide by 60 circuit | ⟶ 1 Hz (one pulse per second)

Fig. 7-7   A 1-second timer system.

INPUT

OUTPUTS

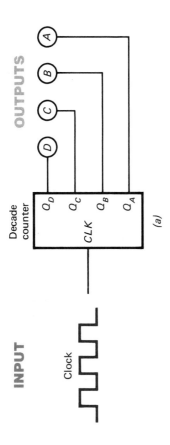

Clock

Decade counter

Q_D Q_C Q_B Q_A

CLK

D C B A

(a)

INPUT   CLK   0 1 2 3 4 5 6 7 8 9 0 1 2 3 4 5 6 7 8 9 0 1 2 3 4 5 6 7 8 9 0 1 2 3 4 5 6 7 8 9

OUTPUT Q_D

(b)

Fig. 7-8 Decade counter used as a divide-by-10 counter. (a) Logic diagram. (b) Waveform diagram.

74192
synchronous
decade up/down
counter

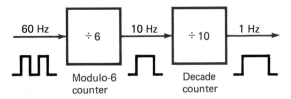

Fig. 7-9 Practical divide-by-60 circuit used as a 1-second timer.

in Fig. 7-10(a); you will see that the 7493 is a ripple counter. You will notice also that the top J-K flip-flop does not have its output (Q_A) connected to the CLK input of the second flip-flop. To operate this counter as a mod-16 counter, you must *externally connect* Q_A to input B on the 7493. The counting sequence is shown in Fig. 7-10(c). The pin diagram for the 7493 counter is drawn in Fig. 7-10(b). To clear or reset the counter to 0000, inputs $R_o(1)$ and $R_o(2)$ must be connected to a logical 1, as shown in Fig. 7-10(d). If these reset inputs are left "floating" (not connected to anything), the counter will not operate. Inputs $R_o(1)$ and $R_o(2)$ will "float high," and therefore the counter remains in the reset condition and will not count.

The second IC counter is the 74192 *synchronous decade up/down counter*. (Signetics again provided the description and diagrams from their data manual.) Figure 7-11 gives some information on the 74192 decade counter. Read the manufacturer's description of the IC counter in Fig. 7-11(a). Notice that the counter has many features. Because the counter is a synchronous counter, its circuitry is quite complex, as seen in Fig. 7-11(b). Figure 7-11(d) diagrams in waveforms some typical sequences you would use on the 74192 counter. You will find upon using this IC that most of the inputs float high. This causes a problem with the CLR input: if it is left unconnected, it floats high and clears the output to 0000.

You probably have already figured that all the features are not used on these IC counters for some applications. Figure 7-12(a) shows the 7493 IC counter being used as a mod-8 counter. Look back at Fig. 7-10 and notice that several inputs and an output are not being used. Figure 7-12(b) shows the 74192 counter as a decade down counter. Six inputs and two outputs are not being used in this circuit. Simplified logic diagrams similar to those in Fig. 7-12 are more common than the complicated diagram in Figs. 7-11(b) or 7-10(a).

Summary

1. Flip-flops are wired together to form binary counters.
2. Counters can operate asynchronously or synchronously. Asynchronous counters are called ripple counters and are simpler to construct than synchronous counters.
3. The modulus of a counter is how many different states it goes through in its counting cycle. A modulo-5 counter counts 000, 001, 010, 011, 100 (0, 1, 2, 3, 4 in decimal).
4. A 4-bit binary counter has four binary place values and counts from 0000 to 1111 (0 to 15 in decimal).

5. Gates can be added to the basic flip-flops in counters to add features. Counters can be made to stop at a certain number. The modulus of a counter can be changed.
6. Counters are designed to count either up or down.
7. Counters are used as frequency dividers. Counters are also widely used to count or sequence events.
8. Manufacturers produce a wide variety of self-contained IC counters.

Questions

1. Draw a logic symbol diagram of a modulo-8 ripple up counter. Use three J-K flip-flops. Show input CLK pulses and three output indicators labeled C, B, and A (C indicator is MSD).

2. Draw a table (similar to Figure 7-1) showing the binary and decimal counting sequence of the mod-8 counter in question 1.

(a) **BLOCK DIAGRAM**

(b) **PIN CONFIGURATION**

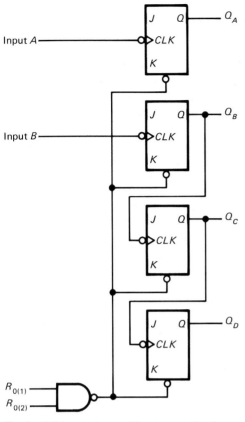

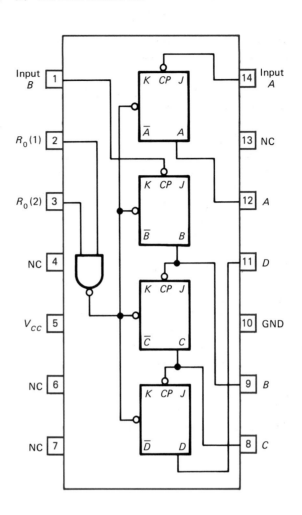

The *J* and *K* inputs shown without connection for reference only and are functionally at a high level.

(c) **COUNT SEQUENCE**

COUNT	OUTPUT			
	Q_D	Q_C	Q_B	Q_A
0	L	L	L	L
1	L	L	L	H
2	L	L	H	L
3	L	L	H	H
4	L	H	L	L
5	L	H	L	H
6	L	H	H	L
7	L	H	H	H
8	H	L	L	L
9	H	L	L	H
10	H	L	H	L
11	H	L	H	H
12	H	H	L	L
13	H	H	L	H
14	H	H	H	L
15	H	H	H	H

Output Q_A is connected to input *B*.

(d) **RESET/COUNT FUNCTION TABLE**

RESET INPUTS		OUTPUT			
$R_0(1)$	$R_0(2)$	Q_D	Q_C	Q_B	Q_A
H	H	L	L	L	L
L	X	Count			
X	L	Count			

Fig. 7-10 4-bit binary counter IC (7493). *(a)* Block diagram. *(b)* Pin configuration. *(c)* Count sequence. *(d)* Reset/count function table. *(Courtesy Signetics.)*

63

(a) DESCRIPTION

This monolithic circuit is a synchronous reversible (up/down) counter having a complexity of 55 equivalent gates. Synchronous operation is provided by having all flip-flops clocked simultaneously so that the outputs change coincidently with each other when so instructed by the steering logic. This mode of operation eliminates the output counting spikes which are normally associated with asynchronous (ripple-clock) counters.

The outputs of the four master-slave flip-flops are triggered by a low-to-high-level transition of either count (clock) input. The direction of counting is determined by which count input is pulsed while the other count input is high.

All four counters are fully programmable; that is, each output may be preset to either level by entering the desired data at the data inputs while the load input is low. The output will change to agree with the data inputs independently of the count pulses. This feature allows the counters to be used as modulo-N dividers by simply modifying the count length with the preset inputs.

A clear input has been provided which forces all outputs to the low level when a high level is applied. The clear function is independent of the count and load inputs. The clear, count, and load inputs are buffered to lower the drive requirements. This reduces the number of clock drivers, etc., required for long words.

These counters were designed to be cascaded without the need for external circuitry. Both borrow and carry outputs are available to cascade both the up- and down-counting functions. The borrow output produces a pulse equal in width to the count-down input when the counter underflows. Similarly, the carry output produces a pulse equal in width to the count-down input when an overflow condition exists. The counters can then be easily cascaded by feeding the borrow and carry outputs to the count-down and count-up inputs respectively of the succeeding counter.

(b) BLOCK DIAGRAM

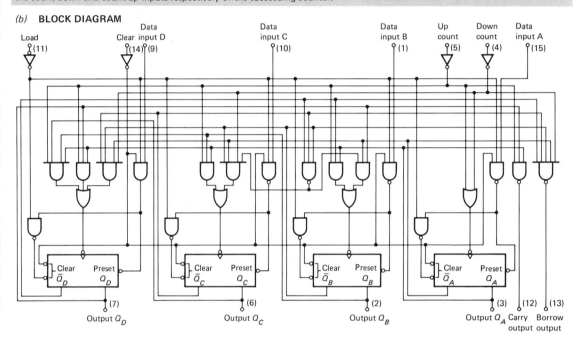

(c) PIN CONFIGURATION

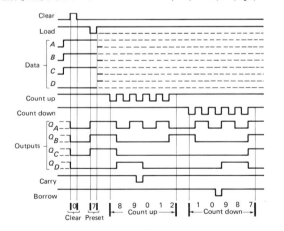

Logic: Low input to load sets $Q_A = A$,
$Q_B = B$, $Q_C = C$, and $Q_D = D$

(d) TYPICAL CLEAR, LOAD, AND COUNT SEQUENCE

Illustrated below is the following sequence:
1. Clear output to zero.
2. Load (preset) to BCD seven.
3. Count up to eight, nine, carry, zero, one, and two.
4. Count down to one, zero, borrow, nine, eight, and seven.

NOTES: A. Clear overrides load, data, and count inputs.
B. When counting up, count-down input must be high; when counting down, count-up input must be high.

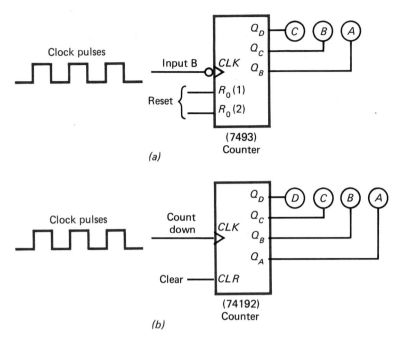

Fig. 7-12 (*a*) 7493 IC wired as a modulus-8 counter. (*b*) 74192
IC wired as a decade down counter.

3. Draw a waveform diagram [similar to Fig. 7-2(*b*)] showing the eight
CLK pulses and the outputs (Q) of FF 1, FF 2, and FF 3 of the mod-8
counter from question 1. Assume you are using negative-edge-
triggered flip-flops.

4. Redesign the mod-8 counter from question 1 to be a mod-5 counter.
Add a 2-input NAND gate to your existing counter. You will need to
use the CLR inputs of the J-K flip-flops.

5. A _____?_____ (asynchronous, synchronous) counter is the most complex circuit.

6. Synchronous counters have the CLK inputs connected in _____?_____
(parallel, series).

7. Draw a logic symbol diagram for a 4-bit ripple down counter. Use
four J-K flip-flops in this mod-16 counter. Show the input CLK
pulses, PS input, and four output indicators labeled D, C, B, and A.

8. If the ripple down counter in question 7 is a recirculating type, what
are the next three counts after 0011, 0010, and 0001?

9. Redesign the 4-bit counter in question 7 to count from binary 1111 to
0000 and then *stop*. Add a 4-input OR gate to your existing circuit to
add this self-stopping feature.

10. Draw a block diagram (similar to Fig. 7-9) showing how you would
use two counters to get an output of 1 Hz with an input of 100 Hz.
Label your diagram.

11. List several places counters are used as frequency dividers.

◁ Fig. 7-11 Synchronous decade up/down counter IC (74192).
(*a*) Description. (*b*) Block diagram. (*c*) Pin configuration. (*d*)
Waveforms. (*Courtesy Signetics.*)

12. Refer to Fig. 7-10 for questions *a* to *f* on the 7493 IC counter:
    a. What is the maximum count length of this counter?
    b. This is a _____?_____ (ripple, synchronous) counter.
    c. What must be the conditions of the reset inputs for the 7493 to count?
    d. This is a(n) _____?_____ (down, up) counter.
    e. The 7493 IC contains _____?_____ (number) flip-flops.
    f. What is the purpose of the NAND gate in the 7493 counter?

13. Refer to Fig. 7-11 for questions *a* to *f* on the 74192 IC counter:
    a. What is the maximum count length of this counter?
    b. This is a _____?_____ (ripple, synchronous) counter.
    c. A logical _____?_____ (0, 1) is needed to clear the counter to 0000.
    d. This is a(n) _____?_____ (down, up, both up and down) counter.
    e. How could we preset the outputs of the 74192 IC to 1111?
    f. How do we get the counter to count downward?

14. Draw a diagram [similar to Fig. 7-12(*a*)] showing how you would wire the 7493 counter as a 4-bit (mod-16) ripple counter. Refer to Fig. 7-10.

15. Draw a diagram [similar to Fig. 7-12(*b*)] showing how you would wire the 74192 counter as a 4-bit (mod-16) synchronous up counter. Refer to Fig. 7-11.

Shift Registers

A typical example of a *shift register* at work is found within a calculator. As you enter each digit on the keyboard, the numbers shift to the left on the display. In other words, to enter the number 268 you do the following. First, you press and release the 2 on the keyboard; a 2 appears on the display. Next, you press and release the 6 on the keyboard; 26 appears on the display. Finally, you press and release the 8 on the keyboard; 268 appears on the display. This example shows two important characteristics of a shift register: (1) it is a *temporary memory* and thus holds the numbers on the display (even if you release the keyboard number), and (2) it shifts the numbers to the left on the display each time you press a new digit on the keyboard. These *memory* and *shifting characteristics* make the shift register extremely valuable in most digital electronic systems. This chapter introduces you to shift registers and explains their operation.

Shift registers are constructed by wiring flip-flops together. We mentioned in Chaps. 6 and 7 that flip-flops had a memory characteristic. This memory characteristic is put to good use in a shift register. Instead of wiring shift registers by using individual gates or flip-flops, you can buy shift registers in IC form.

Shift registers often are used to momentarily store data. Figure 8-1 shows a typical example of where shift registers might be used in a digital system. (We saw the block diagram in Fig. 8-1 before, in Fig. 2-6.) This system could be that of a calculator. Notice the use of shift registers to hold information from the encoder for the processing unit. A shift register is also being employed for temporary storage between the processing unit and the decoder. Shift registers are also used at other locations within a digital system.

8-1 SERIAL LOAD SHIFT REGISTER

A basic shift register is shown in Fig. 8-2. This shift register is constructed from four D flip-flops. This register is called a *4-bit shift register* because it has four places to store data: A, B, C, D.

With the aid of Table 8-1 and Fig. 8-2, let us operate this shift register. First, clear (*CLR* input to 0) all the outputs (A, B, C, D) to 0000. (This situation is shown in line 1, Table 8-1.) Next, set the data and *CLR* inputs to 1 (see line 2, Table 8-1). The outputs remain 0000 while they await a clock pulse. Pulse the *CLK* input once; the output will show 1000 (see line 3, Table 8-1) because the 1 from the D input of FF A was transferred to the Q output on the clock pulse. Now enter 1s on the

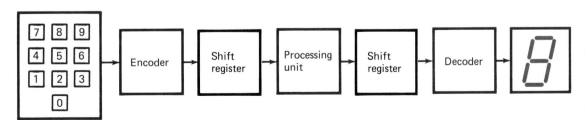

Fig. 8-1 A digital system using shift registers.

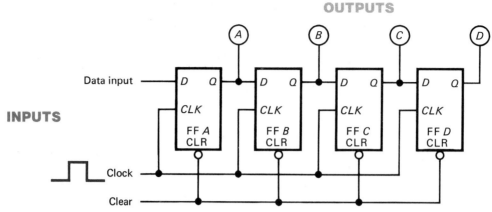

OUTPUTS

Data input

INPUTS

From page 67:
4-bit shift
register

On this page:
Serial load
shift register

Fig. 8-2   4-bit serial load shift register using D flip-flops.

data input (see clock pulses 2 and 3, Table 8-1); these 1s shift across the display to the right.   Next, enter 0s on the data input (see clock pulses 4 to 8, Table 8-1); you can see the 0s being shifted across the display (see lines 6 to 10, Table 8-1).   On clock pulse 9 (see Table 8-1) enter a 1 at the data input.   On pulse 10 the data input is returned to 0.   Pulses 9 to 13 show the single 1 on the display being shifted

to the right.   Line 15 shows the 1 being shifted out the right end of the shift register and being lost.

Remember that the D flip-flop is also called a *delay* flip-flop.   Recall that it simply transfers the data from input D to output Q *after a delay of one clock pulse.*

The circuit diagramed in Fig. 8-2 is referred to as a *serial load shift register*.   The term *se-*

Table 8-1   Operation of a 4-bit serial shift register.

Line number	Clear	Data	Clock pulse number	FF A / A	FF B / B	FF C / C	FF D / D
1	0	0	0	0	0	0	0
2	1	1	0	0	0	0	0
3	1	1	1	1	0	0	0
4	1	1	2	1	1	0	0
5	1	1	3	1	1	1	0
6	1	0	4	0	1	1	1
7	1	0	5	0	0	1	1
8	1	0	6	0	0	0	1
9	1	0	7	0	0	0	0
10	1	0	8	0	0	0	0
11	1	1	9	1	0	0	0
12	1	0	10	0	1	0	0
13	1	0	11	0	0	1	0
14	1	0	12	0	0	0	1
15	1	0	13	0	0	0	0

rial load comes from the fact that only 1 bit of data at a time can be entered in the register. For instance, to enter 0111 in the register, we had to go through the sequence from lines 1 through 6 in Table 8-1. It took five steps (line 2 was not needed) to serially load 0111 into the serial load shift register. To enter 0001 in this serial load shift register we need five steps, as shown in Table 8-1, lines 10 to 14. Another type of loading is called *parallel*, or broadside, loading, in which all the bits of information are loaded at the command of one clock pulse.

The shift register in Fig. 8-2 could become a 5-bit shift register by just adding one more D flip-flop. Shift registers typically come in 4-, 5-, and 8-bit sizes. Shift registers also can be wired using other flip-flops. J-K flip-flops and clocked R-S flip-flops are also used to wire shift registers.

8-2 PARALLEL LOAD SHIFT REGISTER

The serial load shift register we studied in the last unit has two disadvantages: it permits only 1 bit of information to be entered at a time, and it loses all its data out the right side when it shifts right. Figure 8-3(*a*) illustrates a system that permits *parallel loading* of 4 bits at once. These inputs are the data inputs A, B, C, and D on Fig. 8-3. This system could also incorporate a *recirculating* feature that would put the output data back into the input so it would not be lost.

An actual wiring diagram of the *4-bit parallel load recirculating shift register* is drawn in Fig. 8-3(*b*). This shift register uses four J-K flip-flops. Notice the recirculating lines leading from the Q and Q̄ outputs of FF D back to the J and K inputs of FF A. These feedback

Recirculating feature

4-bit parallel load recirculating shift register

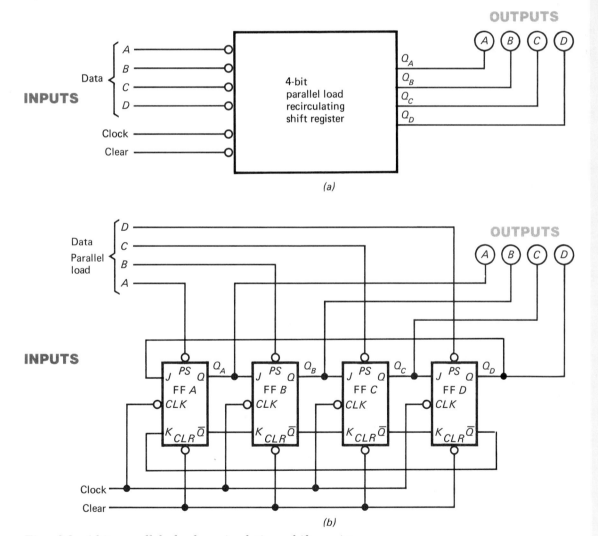

Fig. 8-3 4-bit parallel load recirculating shift register. (*a*) Block diagram. (*b*) Wiring diagram.

69

lines cause the data that would normally be lost out of FF *D* to recirculate through the shift register. The *CLR* input erases the outputs to 0000 when enabled by a logical 0. The parallel load data inputs *A*, *B*, *C*, and *D* are connected to the preset (*PS*) inputs of the flip-flops to set 1s at any output position (*A*, *B*, *C*, *D*). If the switches attached to the parallel load data inputs are even temporarily switched to a 0, that output will be preset to a logical 1. The clock pulsing the *CLK* inputs of the J-K flip-flops will cause data to be shifted to the right. The data from FF *D* will be recirculated back to FF *A*.

Table 8-2 will help you understand the operation of the parallel load shift register. As you turn on the power, the outputs may take any combination, such as the one in line 1. Line 2 shows the register being erased with the *CLR* input. Line 3 shows 0100 being loaded into the register using the parallel load data switches. Lines 4 to 8 show five clock pulses and the shifting of the data to the right. Look at lines 5 and 6: the 1 is being recirculated from the right end (FF *D*) of the register back

to the left end (FF *A*). We say the 1 is being recirculated.

Line 9 shows the register being erased again by the *CLR* input. New information (0110) is being loaded in the data inputs in line 10. Lines 11 to 15 illustrate the register being shifted five times by clock pulses. Note that it takes four clock pulses to come back to the original data in the register (compare lines 11 and 15 or lines 4 and 8 in Table 8-2). The recirculating feature of the shift register in Fig. 8-3(*b*) can be disconnected by taking out the recirculating lines. The register would then be a parallel loaded shift register without the recirculating feature.

8-3 A UNIVERSAL SHIFT REGISTER

When reviewing data manuals you will see that manufacturers produce many shift registers in IC form. In this section one such IC shift register will be studied: the *74194 4-bit bidirectional universal shift register*.

The 74194 IC is a very adaptable shift regis-

Table 8-2   Operation of a 4-bit parallel load recirculating shift register.

Line number	Clear	Parallel load data				Clock pulse number	FF *A*	FF *B*	FF *C*	FF *D*
		A	*B*	*C*	*D*		*A*	*B*	*C*	*D*
1	1	1	1	1	1	0	1	1	1	0
2	0	1	1	1	1	0	0	0	0	0
3	1	1	0	1	1	0	0	1	0	0
4	1	1	1	1	1	1	0	0	1	0
5	1	1	1	1	1	2	0	0	0	1
6	1	1	1	1	1	3	1	0	0	0
7	1	1	1	1	1	4	0	1	0	0
8	1	1	1	1	1	5	0	0	1	0
9	0	1	1	1	1		0	0	0	0
10	1	1	0	0	1		0	1	1	0
11	1	1	1	1	1	6	0	0	1	1
12	1	1	1	1	1	7	1	0	0	1
13	1	1	1	1	1	8	1	1	0	0
14	1	1	1	1	1	9	0	1	1	0
15	1	1	1	1	1	10	0	0	1	1

(a) DESCRIPTION

This bidirectional shift register is designed to incorporate virtually all of the features a system designer may want in a shift register. The circuit contains 45 equivalent gates and features parallel inputs, parallel outputs, right-shift and left-shift serial inputs, operating-mode-control inputs, and a direct overriding clear line. The register has four distinct modes of operation, namely:

    Parallel (broadside) load
    Shift right (in the direction Q_A toward Q_D)
    Shift left (in the direction Q_D toward Q_A)
    Inhibit clock (do nothing)

Synchronous parallel loading is accomplished by applying the four bits of data and taking both mode control inputs, $S0$ and $S1$, high. The data are loaded into the associated flip-flops and appear at the outputs after the positive transition of the clock input. During loading, serial data flow is inhibited.

Shift right is accomplished synchronously with the rising edge of the clock pulse when $S0$ is high and $S1$ is low. Serial data for this mode is entered at the shift-right data input. When $S0$ is low and $S1$ is high, data shifts left synchronously and new data is entered at the shift-left serial input.

Clocking of the flip-flop is inhibited when both mode control inputs are low. The mode controls of the S54194/N74194 should be changed only while the clock input is high.

(b) BLOCK DIAGRAM

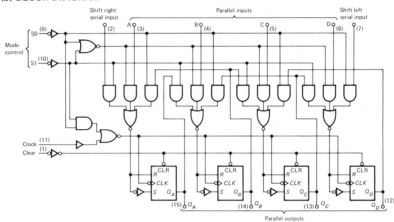

(c) PIN CONFIGURATION

(d) TRUTH TABLE

INPUTS										OUTPUTS			
	MODE			SERIAL		PARALLEL				Q_A	Q_B	Q_C	Q_D
CLEAR	$S1$	$S2$	CLOCK	LEFT	RIGHT	A	B	C	D				
L	X	X	X	X	X	X	X	X	X	L	L	L	L
H	X	X	L	X	X	X	X	X	X	Q_{A0}	Q_{B0}	Q_{C0}	Q_{D0}
H	H	H	↑	X	X	a	b	c	d	a	b	c	d
H	L	H	↑	X	H	X	X	X	X	H	Q_{An}	Q_{Bn}	Q_{Cn}
H	L	H	↑	X	L	X	X	X	X	L	Q_{An}	Q_{Bn}	Q_{Cn}
H	H	L	↑	H	X	X	X	X	X	Q_{Bn}	Q_{Cn}	Q_{Dn}	H
H	H	L	↑	L	X	X	X	X	X	Q_{Bn}	Q_{Cn}	Q_{Dn}	L
H	L	L	X	X	X	X	X	X	X	Q_{A0}	Q_{B0}	Q_{C0}	Q_{D0}

H = high level (steady state)
L = low level (steady state)
X = irrelevant (any input, including transitions)
↑ = transition from low to high level
$a,b,c,d,$ = the level of steady state input at inputs $A,B,C,$ or D, respectively
$Q_{A0}, Q_{B0}, Q_{C0}, Q_{D0}$ = the level of Q_A, Q_B, Q_C, Q_D, respectively, before the indicated steady state input conditions were established
$Q_{An}, Q_{Bn}, Q_{Cn}, Q_{Dn}$ = the level of Q_A, Q_B, Q_C, Q_D, respectively, before the most recent ↑ transition of the clock

(e) TYPICAL CLEAR, SHIFT, AND LOAD SEQUENCES

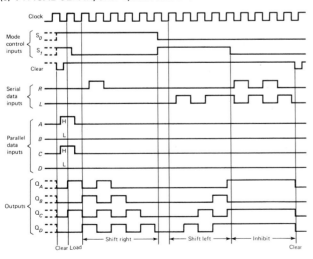

Fig. 8-4 4-bit universal shift register (74194). (*a*) Description. (*b*) Block diagram. (*c*) Pin configuration. (*d*) Truth table. (*e*) Waveforms. (*Courtesy Signetics.*)

ter and has most of the features we have seen so far in one IC package. In 74194 IC register can shift right or left. It can be loaded serially or in parallel. Several 4-bit 74194 IC registers can be cascaded to make an 8-bit or longer shift register. And this register can be made to recirculate data.

The Signetics data manual contains the descriptions, diagrams, and tables shown in Fig. 8-4. Read the description of the 74194 shift register in Fig. 8-4(a) for a good overview of what this shift register can do.

A logic diagram of the 74194 register is reproduced in Fig. 8-4(b). Because it is a 4-bit register, the circuit contains four flip-flops.

Extra gating circuitry is needed for the many features of this universal shift register. The pin configuration in Fig. 8-4(c) will help you determine the terms attached to each input and output. Of course, the pin diagram is also a must when actually wiring a 74194 IC.

The truth table and waveform diagrams in Fig. 8-4(d) and (e) are very helpful in determining exactly how the 74194 IC register works because they illustrate the clear, load, shift right, shift left, and inhibit modes of operation. As you use the 74194 universal shift register you will have occasion to look quite carefully at the truth table and waveform diagrams.

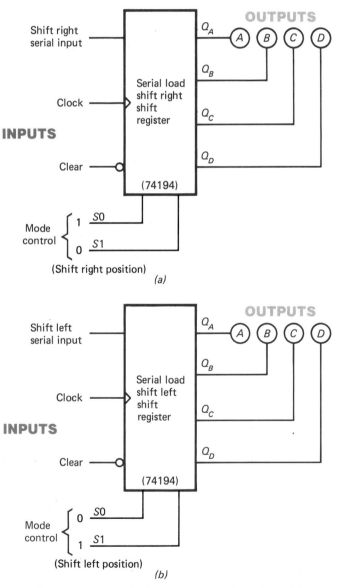

Fig. 8-5 (a) 74194 IC wired as a 4-bit serial load shift-right register. (b) 74194 IC wired as a 4-bit serial load shift-left register.

8-4 USING THE 74194 IC SHIFT REGISTER

In this section we shall use the 74194 universal shift register in several ways. Figure 8-5(a) and (b) show the 74194 IC being used as serial loaded registers. A *serial load shift-right register* is shown in Fig. 8-5(a). This register operates exactly like the serial shift register in Fig. 8-2. Table 8-1 could also be used to chart the performance of this new shift register. Notice that the mode control inputs ($S0$, $S1$) must be in the positions shown for the 74194 IC to operate in its shift-right mode. Shifting to the right is defined by the manufacturer as shifting from Q_A to Q_D. The register in Fig. 8-5(a) shifts data to the right, and as it leaves Q_D it is lost.

The 74194 IC has been rewired slightly in Fig. 8-5(b). The shift-left serial input is used, and the mode control inputs *have been changed*, as shown. This register enters data at D (Q_D) and shifts it toward A (Q_A) with each pulse of the clock. This register is a *serial load shift-left register*. This is not a recirculating-type register as wired in Fig. 8-5(b).

In Fig. 8-6 the 74194 IC has been wired as a *parallel load shift-right/left register*. With a single clock pulse the data from the parallel load inputs A, B, C, and D will appear on the display. The loading will happen only when the mode controls ($S0$, $S1$) are set at 1, as

shown. The mode control can then be changed to one of three types of operations: shift right, shift left, or inhibit. The shift-right and shift-left serial inputs both are connected to 0 to feed in 0s to the register in the shift-right or shift-left modes of operation. With the mode control in the inhibit position ($S0 = 0$, $S1 = 0$), the data does not shift right or left but stays in position in the register. When using the 74194 IC you must remember the mode control inputs because they control the operation of the entire register. The *CLR* input erases the register to 0000 when enabled by a 0. The *CLR* input overrides all other inputs.

Two 74194 IC shift registers are connected in Fig. 8-7 to form an *8-bit parallel load shift-right register*. The *CLR* input erases the outputs to 0000 0000. The parallel load inputs A to H allow entry of all 8 bits of data on a single clock pulse (mode control–$S0 = 1$, $S1 = 1$). With the mode control to the shift-right position ($S0 = 1$, $S = 0$), the register shifts right for each clock pulse. Notice that a recirculating line has been placed from output H (output Q_D of register 2) back to the shift-right serial input of shift register 1. Data that normally would be lost out of output H is recirculated back to position A in the register. Both inputs $S0$ and $S1$ at 0 will inhibit the shifting of data in the shift register.

As you have just seen, the 74194 IC 4-bit bi-

Serial load shift-right register

Serial load shift-left register

Parallel load shift-right/left register

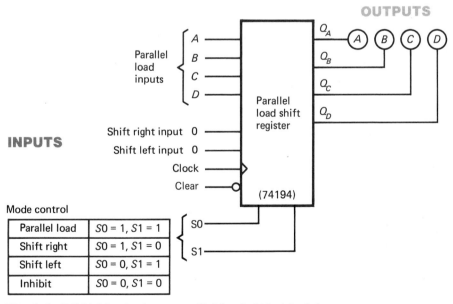

INPUTS

OUTPUTS

Mode control

Parallel load	$S0 = 1, S1 = 1$
Shift right	$S0 = 1, S1 = 0$
Shift left	$S0 = 0, S1 = 1$
Inhibit	$S0 = 0, S1 = 0$

Fig. 8-6 74194 IC wired as a parallel load shift-right/left register.

Temporary
memory

Delay lines

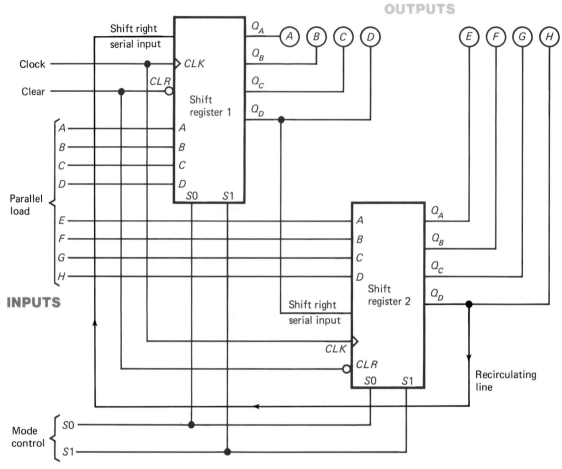

OUTPUTS

INPUTS

Fig. 8-7   Two 74194 ICs wired as an 8-bit parallel load shift-right register.

directional universal shift register is very useful. The circuits in this unit are only a few examples of how the 74194 IC can be used. Remember that all shift registers use as their basis the memory characteristic of a flip-flop. Shift registers often are used as temporary memories. Shift registers also can be used to convert serial data to parallel data or parallel data to serial data. And shift registers can be used to delay information (delay lines). Shift registers are also used in some arithmetic circuits.

Summary

1. Flip-flops are wired together to form shift registers.
2. A shift register has both a memory and a shift characteristic.
3. A serial load shift register is one that permits only 1 bit of data to be entered per clock pulse.
4. A parallel load shift register is one that permits all data bits to be entered at one time.

5. A register that is recirculating feeds output data back into the input.
6. Shift registers are designed to shift either left or right.
7. Manufacturers produce many adaptable universal shift registers.
8. Shift registers are widely used as temporary memories and for shifting data. They also have other uses in digital electronic systems.

Questions

1. Draw a logic symbol diagram of a 5-bit serial load shift-right register. Use five D flip-flops. Label inputs as data, *CLK*, and *CLR*. Label outputs as A, B, C, D, and E. The circuit will be similar to the one in Fig. 8-2.

2. Explain how you would clear to 00000 the 5-bit register you drew in question 1.

3. After clearing the 5-bit register, explain how you would enter (load) 10000 into the register you drew in question 1.

4. After clearing the 5-bit register, explain how you would enter (load) 00111 into the register you drew in question 1.

5. Refer to the register you drew in question 1. List the contents of the register after each clock pulse shown in *b* to *e* (assume data input = 0):
   a. Original output = 01001 ($A = 0, B = 1, C = 0, D = 0, E = 1$)
   b. After 1 clock pulse =
   c. After 2 clock pulses =
   d. After 3 clock pulses =
   e. After 4 clock pulses =

6. Refer to Table 8-1. Assume the data input was left at 1 in lines 12 to 15. List the outputs that would be recorded on the table under these conditions on this serial shift-right register:
   a. Output in line 12 =
   b. Output in line 13 =
   c. Output in line 14 =
   d. Output in line 15 =

7. Refer to Table 8-2. The parallel load register using J-K flip-flops [Fig. 8-3(*b*)] needs _____?_____ (0, 1, 3, 4) clock pulse(s) to load data from the data inputs.

8. Refer to Fig. 8-6. The parallel load register using the 74194 IC needs _____?_____ (0, 1, 3, 4) clock pulse(s) to load data from the parallel load inputs.

9. A _____?_____ (serial, parallel) load shift register is the simplest circuit to wire.

10. A _____?_____ (serial, parallel) load shift register is the easiest to load.

11. List several uses of shift registers in digital systems.

12. Refer to Fig. 8-4 for questions *a* to *i* on the 74194 IC shift register:
    a. How many bits of information can this register hold?
    b. List the four modes of operation for this register.
    c. What is the purpose of the Mode Control inputs (S0, S1)?
    d. The _____?_____ input will override all other inputs on this register.
    e. What type of flip-flops and how many are used in this shift register?
    f. The register shifts on the _____?_____ (negative-, positive-) going edge of the clock pulse.
    g. What does the inhibit mode of operation mean?
    h. By definition, to shift left means to shift data from _____?_____ to _____?_____ (use letters).
    i. This register can be loaded _____?_____ (serially, in parallel, both serially and in parallel).

Arithmetic Circuits

■ The public's imagination has been captured by computers and modern-day calculators, probably because these machines perform human arithmetical tasks with such fantastic speed and accuracy. This chapter deals with some logic circuits that can add and subtract. (Of course, the adding and subtracting will be done in binary.) Regular logic gates will be wired together to form *adders* and *subtractors*.

9-1 BINARY ADDITION

Remember that in a binary number such as 101011, the leftmost 1 is the most significant bit (MSB), and the right digit (1) is the least significant bit (LSB). Also remember the place values given to the binary number: 1s, 2s, 4s, 8s, 16s, and 32s.

You probably still recall learning your addition and subtraction tables when you were in grade school. This was a difficult task for the decimal number system because it has so many combinations. This section deals with the simple task of adding numbers in binary. Because they have only two digits (0 and 1), the binary addition tables are simple. Figure

9-1(a) shows the binary addition tables. Just as is the case of adding with decimals, the first three problems are easy. The next problem is 1 + 1. In decimal that would be 2. In binary a 2 is written as 10 (say "one-zero"). Therefore, in binary 1 + 1 = 0, with a carry of 1 to the next most significant place value.

Figure 9-1(b) shows some examples of adding numbers in binary. The problems are also shown in decimal so you can check your understanding of binary addition. The first problem is adding binary 101 to 10, which equals 111 (decimal 7). This problem was simple using the addition tables in Figure 9-1(a). The second problem in Fig. 9-1(b) is adding binary 1010 to 11. Here you must notice

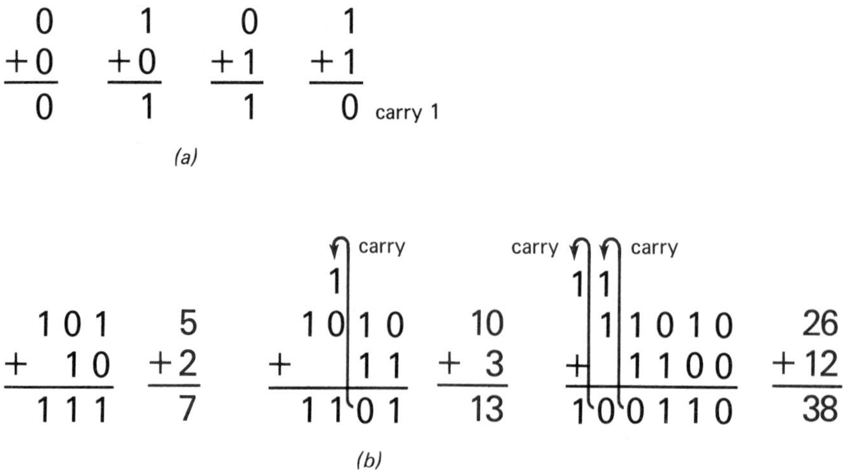

```
  0     1     0     1
 +0    +0    +1    +1
 ──    ──    ──    ──
  0     1     1     0  carry 1
```

(a)

```
              ┌─ carry        carry ─┐ ┌─ carry
              1                    1 1
   1 0 1    5     1 0 1 0   10      1 1 0 1 0   26
 +   1 0   +2   +     1 1  + 3    +   1 1 0 0  +12
 ──────    ──   ─────────   ──    ───────────  ──
   1 1 1    7     1 1 0 1   13    1 0 0 1 1 0   38
```

(b)

Fig. 9-1   (a) Binary addition tables.   (b) Sample binary addition problems.

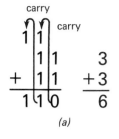

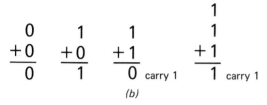

Fig. 9-2 (a) Sample binary addition problem. (b) Short-form binary addition table.

that a 1 + 1 = 0 plus a carry from the 2s place to the 4s place, as shown in the diagram. The answer to this problem is 1101 (decimal 13). In the third problem in Fig. 9-1(b) the binary number 11010 is added to 1100. In the figure note two carries with the solution as 100110 (decimal 38).

Another sample addition problem is shown in Fig. 9-2(a). The solution looks simple until we get to the 2s column and find 1 + 1 + 1 in binary. This equals 3 in decimal, which is 11 (say "one-one") in binary. This is one situation we left out of the first group of binary addition tables. Looking carefully at Fig. 9-2, you will see that the 1 + 1 + 1 situation can arise in any column except the 1s column. So the binary addition table in Fig. 9-1(a) is correct for the *1s column only*. The new short-form addition table in Fig. 9-2(b) adds the other possible combination of 1 + 1 + 1. The addition table in Fig. 9-2(b), then, is for all the place values (2s, 4s, 8s, 16s, and so on) except the 1s column.

To be an intelligent worker on digital equipment you must master binary addition. Several practice problems are provided in the first self-test.

9-2 HALF ADDER

The addition table in Fig. 9-1(a) may be thought of as a truth table. The input side of the truth table would be the numbers being added. In Fig. 9-3(a) this would be the A and B input columns. The truth table would

need *two* output columns, one column for the sum and one column for the carry. The sum column is labeled with the summation symbol (Σ). The carry column is labeled with a C_o. The C_o stands for carry output or *carry out*. A convenient block symbol for the adder that will perform the job of the truth table is shown in Fig. 9-3(b). This circuit is called a *half-adder* circuit. The half-adder circuit has two inputs (A, B) and two outputs (Σ, C_o).

Take a careful look at the half-adder truth table in Fig. 9-3(a). What is the Boolean expression needed for the C_o output? The Boolean expression is $A \cdot B = C_o$. You need a 2-input AND gate to take care of output C_o.

Now what is the Boolean expression for the sum (Σ) output of the half adder in Fig. 9-3(a). The Boolean expression is $\overline{A} \cdot B + A \cdot \overline{B} = \Sigma$. We could use two AND gates and one OR gate to do the job. If you look closer you will notice that this pattern is also that of an exclusive OR (XOR) gate. The simplified Boolean expression is then $A \oplus B = \Sigma$. In other words, we find that only one 2-input XOR gate is needed to produce the sum output.

Using a 2-input AND gate and a 2-input XOR gate, a logic symbol diagram for a half adder is drawn in Fig. 9-4. The half-adder circuit will add only the LSB column (1s column) in a binary addition problem. A circuit

From page 76:
Binary addition

On this page:
Half-adder

Half-adder truth table

TRUTH TABLE

INPUTS		OUTPUTS	
B	**A**	Σ	C_o
0	0	0	0
0	1	1	0
1	0	1	0
1	1	0	1
Binary digits to be added		Sum	Carry out
		XOR	AND

(a)

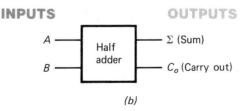

(b)

Fig. 9-3 Half adder. (a) Truth table. (b) Block symbol.

Half adder

INPUTS   **OUTPUTS**

A ——————————
                    XOR ———— Σ (Sum)
B ——————————

                    AND ———— C_o (Carry out)

Fig. 9-4 Logic diagram for a half adder.

called a *full adder* must be used for the 2s, 4s, 8s, 16s, and so forth places in binary addition.

9-3 FULL ADDER

Figure 9-2(b) was a short form of the binary addition table, with the 1 + 1 + 1 situation

shown. The truth table in Fig. 9-5(a) shows all the possible combinations of A, B, and C_{in} (carry in). This truth table is for a full adder. Full adders are used for all binary place values except the 1s place. The full adder must be used when it is possible to have an extra *carry input* (C_{in}). A block diagram of a full adder is shown in Fig. 9-5(b). The full adder has three inputs: C_{in}, A, and B. These three inputs must be added to get the Σ and C_o outputs.

One of the easiest methods of forming the combinational logic for a full adder is diagramed in Fig. 9-5(c); two half-adder circuits and an OR gate are used. The expression for this arrangement is $A \oplus B \oplus C = \Sigma$ (Σ for the sum). The expression for the carry out is $A \cdot B + C_{in} \cdot (A \oplus B) = C_o$. The logic circuit in Fig. 9-6(a) is a full adder. This circuit

TRUTH TABLE

	INPUTS		OUTPUTS	
C_{in}	B	A	Σ	C_o
0	0	0	0	0
0	0	1	1	0
0	1	0	1	0
0	1	1	0	1
1	0	0	1	0
1	0	1	0	1
1	1	0	0	1
1	1	1	1	1
Carry + B + A			Sum	Carry out

(a)

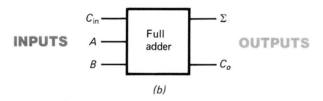

INPUTS   **OUTPUTS**

(b)

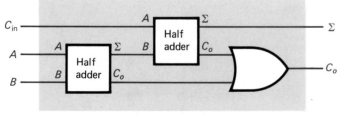

(c)

Fig. 9-5 Full adder. (a) Truth table. (b) Block symbol. (c) Constructed from half adders and an OR gate.

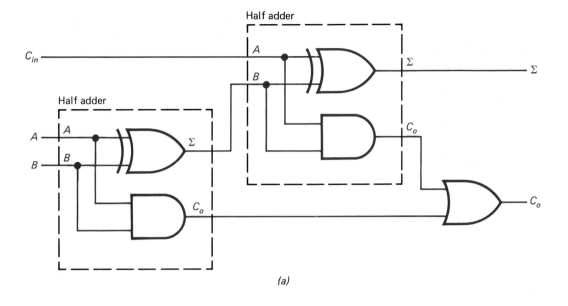

(a)

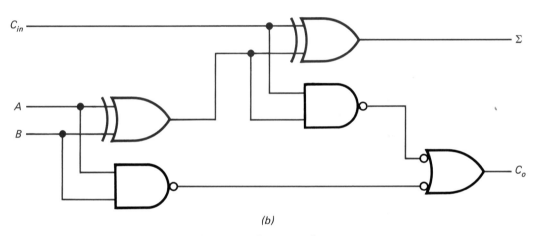

(b)

Fig. 9-6 Full adder. (*a*) Logic diagram. (*b*) Logic diagram using **XOR** and **NAND** gates.

is based upon the block diagram using two half adders, as shown in Fig. 9-5(*c*). Directly below this logic diagram is a somewhat easier logic circuit to wire. Figure 9-6(*b*) contains two XOR gates and three NAND gates, which makes the circuit fairly easy to wire. Notice that the circuit in Fig. 9-6(*b*) is exactly the same as the one in Fig. 9-6(*a*), except NAND gates have been substituted for AND and OR gates.

Half and full adders are used together. For the problem in Fig. 9-2(*a*) we would need one half adder for the 1s place and two full adders for the 2s place and the 4s place value. Half and full adders are rather simple circuits. However, many of these circuits are needed to add longer problems.

Self Test

Check your understanding by answering questions 1 to 6.

1. Do the binary addition problems in *a* to *j* (check yourself by using decimal addition):

a.	1010 + 100	*f.*	10 + 11
b.	111 + 100	*g.*	11 + 1
c.	1010 + 111	*h.*	1111 + 1001
d.	1110 + 101	*i.*	10011 + 111
e.	100 + 11	*j.*	1100 + 1100

79

2. Draw a block diagram of a half adder. Label inputs as A and B; label outputs as Σ and C_o.

3. Draw a truth table for a half adder (inputs A, B; outputs Σ, C_o).

4. Draw a logic symbol diagram of a half adder.

5. Draw a block diagram of a full adder. Label inputs A, B, and C_{in}; label outputs Σ and C_o.

6. Draw a truth table for a full adder (inputs A, B, C_{in}; outputs Σ, C_o).

9-4 3-BIT ADDER

Half and full adders are connected to form adders that will add several binary digits (bits) at one time. The system in Fig. 9-7 adds two 3-bit numbers. The numbers being added are represented by $A_2A_1A_0$ and $B_2B_1B_0$. Numbers from the 1s place value column are entered into the 1s adder or half adder. The inputs to the 2s adder are the carry from the half adder and the new bits A_1 and B_1 from the problem. The 4s bit adder adds A_2 and B_2 and carry in (from 2s adder). The total sum is shown in binary in the lower right. The output also has an 8s place value to take care of any binary number over 111 in the sum. Notice that the

4s adder's output (C_o) is connected to the 8s sum indicator.

The 3-bit binary adder is organized as you would *add and carry* in hand-done arithmetic. The electronic adder in Fig. 9-7 is just very much faster than doing the same problem by hand. You noticed that multibit adders use a half adder for the 1s column only; all other bits use a full adder. This type of adder is called a *parallel adder*.

9-5 BINARY SUBTRACTION

You will find that *adders* and *subtractors* are very similar. You use *half subtractors* and *full subtractors* just as you use half and full adders. Binary subtraction tables are shown in Fig. 9-8(a). Converting these rules into truth table form gives the table in Fig. 9-8(b). On the input side, B is subtracted from A to give output Di (difference). If B is larger than A, such as in line 2, we need a *borrow*, which is shown in the column labeled B_0 (borrow out).

A block diagram of a half subtractor is shown in Fig. 9-9(a). Inputs A and B are on the left. Outputs Di and B_o are on the right side of the diagram. Looking at the truth table in Fig. 9-8(b), we can determine the Boo-

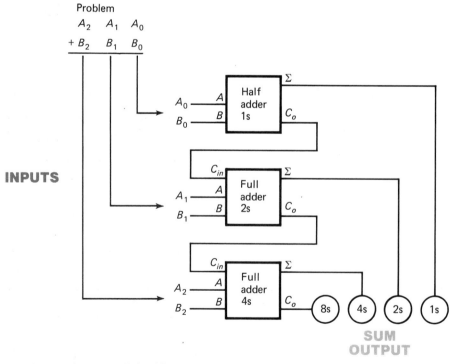

Fig. 9-7 A 3-bit parallel adder.

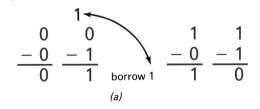

(a)

TRUTH TABLE

INPUTS		OUTPUTS	
A	**B**	**Di**	**B_o**
0	0	0	0
0	1	1	1
1	0	1	0
1	1	0	0
A - B		Difference	Borrow

(b)

Fig. 9-8 (a) **Binary subtraction tables.** (b) **Truth table for the half subtractor.**

lean expressions for the half subtractor. The expression for the Di column is $A \oplus B = Di$. That is the same as for the half adder [see Fig. 9-3(a)]. The Boolean expression for the B_o column will be $\overline{A} \cdot B = B_o$. Combining these two expressions in a logic diagram gives the logic circuit in Fig. 9-9(b). This is the logic circuit for a half subtractor; notice how much it looks like the half-adder circuit in Fig. 9-4.

When you subtract several columns of bi-

INPUTS   OUTPUTS

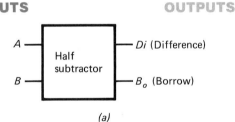

(a)

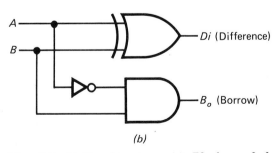

(b)

Fig. 9-9 Half subtractor. (a) **Block symbol.** (b) **Logic diagram.**

nary digits, you must take into account the borrowing. Suppose you were subtracting the numbers in Fig. 9-10(a). You might keep track of the differences and borrows as shown in the figure. Look over the subtraction problem carefully, and check if you can do binary subtraction by this longhand method. (You can check yourself on the next self test.)

A truth table that considers all the possible combinations in binary subtraction is in Fig. 9-10(b). For instance, line 5 of the table is the situation in the 1s column in Fig. 9-10(a). The 2s column equals line 3, the 4s column line 6, the 8s column line 3, the 16s column line 2, and the 32s column line 6 of the truth table.

A block diagram of a full subtractor is drawn in Fig. 9-11(a). The inputs A, B, and B_{in} are on the left; the outputs Di and B_o are on the right. Like the full adder, the full subtractor can be wired using two half subtractors and an OR gate. Figure 9-11(b) is a full subtractor showing how half subtractors are used. An

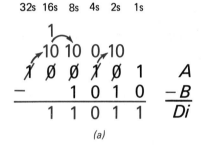

(a)

TRUTH TABLE

INPUTS			OUTPUTS	
A	**B**	**B_{in}**	**Di**	**B_o**
0	0	0	0	0
0	0	1	1	1
0	1	0	1	1
0	1	1	0	1
1	0	0	1	0
1	0	1	0	0
1	1	0	0	0
1	1	1	1	1
A - B - B_{in}			Difference	Borrow out

(b)

Fig. 9-10 (a) **Sample binary subtraction problem.** (b) **Truth table for a full subtractor.**

81

4-bit parallel subtractor

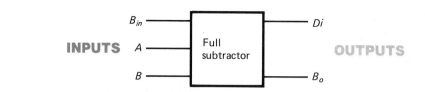

(a)

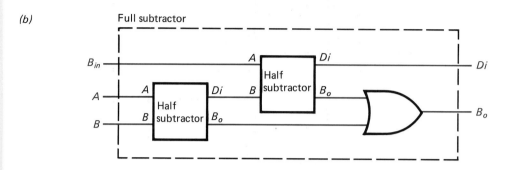

(b)

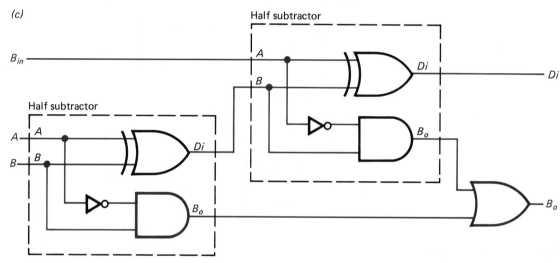

(c)

Fig. 9-11   Full subtractor.   (*a*) Block symbol.   (*b*) Constructed with half subtractors and an OR gate.   (*c*) Logic diagram.

actual logic diagram for a full subtractor is shown in Fig. 9-11(*c*). This circuit performs as a full subtractor as specified in the truth table in Fig. 9-10(*b*). The AND-OR circuit on the B_o output can be converted to three NAND gates if you want. The circuit would then be similar to the full adder circuit in Fig. 9-6(*b*).

9-6 PARALLEL SUBTRACTOR

Half and full subtractors are wired together to perform as a *parallel subtractor*. You already saw adders connected as parallel adders. An example of a parallel adder is the 3-bit adder in Fig. 9-7. A parallel subtractor is wired in a

similar manner. The adder in Fig. 9-7 is considered a parallel adder because all the digits from the problem flow into the adder at the same time.

Figure 9-12 diagrams the wiring of a single half subtractor and three full subtractors. This forms a 4-bit parallel subtractor that will subtract binary number $B_3B_2B_1B_0$ from binary number $A_3A_2A_1A_0$. Notice that the top subtractor (half subtractor) subtracts the LSBs (1s place). The B_0 output of the 1s subtractor is tied to the 2s subtractor. Each subtractor's B_o output is connected to the next more significant digit's borrow input. These "borrow" lines keep track of the borrows we discussed earlier.

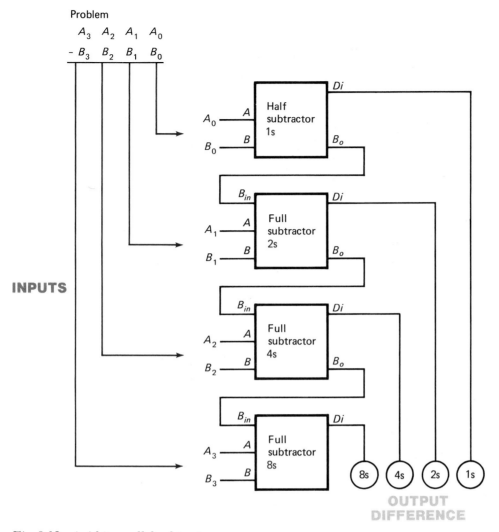

Problem

$$A_3 \quad A_2 \quad A_1 \quad A_0$$
$$- B_3 \quad B_2 \quad B_1 \quad B_0$$

INPUTS

1s complement

End-around carry

Fig. 9-12 A 4-bit parallel subtractor.

9-7 USING ADDERS FOR SUBTRACTION

In Secs. 9-1 to 9-6 we found that there are circuits for adding or subtracting binary numbers. To simplify circuitry in a calculating machine, it would be convenient to have a more universal device for calculations. With a few little tricks we can use an adder to also do subtraction.

There is a mathematical technique that will help us use an adder to do binary subtraction. The technique is outlined in Fig. 9-13. The problem is to subtract decimal number 6 from 10 (binary: $1010 - 0110$). The subtracting is shown being done first in decimal, then in binary, and finally by the special technique. In the special technique the steps are

first to write the *1s complement* of the number being subtracted (change all 1s to 0s and all 0s to 1s), and then add. The 1s complement of 0110 is 1001, as shown. The temporary answer to this addition is shown as 10011. Next, the last carry on the left is carried around to the 1s place (see the arrow on the diagram). This is called an *end-around carry*. When the end-around carry is added to the rest of the number, the result is the *difference* between the original binary numbers 1010 and 0110. The answer to this problem is shown as binary 100 (decimal 4).

Using the 1s complement and end-around carry method is somewhat difficult in longhand. However, this method is quite easy to do with logic circuits. You can see that this method uses an adder to do subtraction. You

83

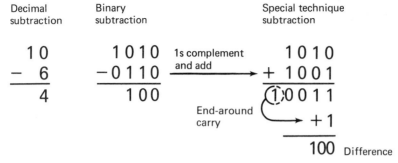

Decimal subtraction

$$\begin{array}{r} 10 \\ -\ 6 \\ \hline 4 \end{array}$$

Binary subtraction

$$\begin{array}{r} 1010 \\ -0110 \\ \hline 100 \end{array}$$

Special technique subtraction

1s complement and add →

$$\begin{array}{r} 1010 \\ +\ 1001 \\ \hline (1)0011 \end{array}$$

End-around carry

→ +1

$$\overline{\qquad 100} \quad \text{Difference}$$

**1s complement
and end-around
carry subtraction**

Fig. 9-13 An example of 1s complement and end-around carry subtraction.

should know how to do 1s complement and end-around carry subtraction. (There are a few practice problems in the next self test.)

Now let us use adders to do binary subtraction. Figure 9-14 shows four *full adders* (labeled FA) being used to perform binary subtraction. Pay special attention to the four

inverters that complement the binary number represented by $B_3B_2B_1B_0$. The inverters give the 1s complement input at the B inputs to each full adder. The adder adds the binary numbers represented by $A_3A_2A_1A_0$ and $\overline{B}_3\overline{B}_2\overline{B}_1\overline{B}_0$. The extra carry at C_o of the 8s full adder is carried back to the 1s adder by the

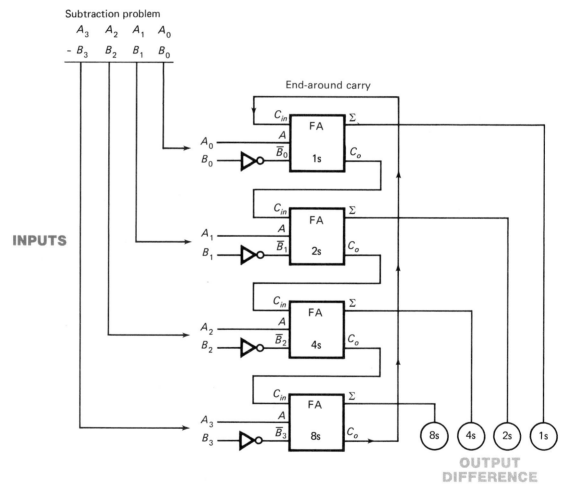

Fig. 9-14 Using full adders and inverters to construct a 4-bit binary subtractor.

end-around carry line shown. The indicators at the lower right show the *difference* between the binary numbers $A_3A_2A_1A_0$ and $B_3B_2B_1B_0$.

Self Test

Check your understanding by answering questions 7 to 16.

7. Do the binary subtraction problems in *a* to *j* (check yourself using decimal subtraction):

a.	11 – 10	*f.*	1010 – 101	
b.	100 – 10	*g.*	10010 – 11	
c.	111 – 111	*h.*	1001 – 110	
d.	1111 – 100	*i.*	1110 – 111	
e.	100 – 11	*j.*	1000 – 01	

8. Do the binary subtraction problems in *a* to *j* using the 1s complement and end-around carry method (show your work):
   - *a.* 11 – 10 = ___?___
   - *b.* 111 – 010 = ___?___
   - *c.* 1000 – 0111 = ___?___
   - *d.* 110 – 100 = ___?___
   - *e.* 1001 – 0111 = ___?___
   - *f.* 1011 – 0110 = ___?___
   - *g.* 101 – 011 = ___?___
   - *h.* 1010 – 0011 = ___?___
   - *i.* 1100 – 0110 = ___?___
   - *j.* 1101 – 0010 = ___?___

9. Draw a block diagram of a half subtractor. Label inputs A and B; label outputs Di and B_o.

10. Draw a truth table for a half subtractor (inputs A, B; outputs Di, B_o).

11. Draw a block diagram of a full subtractor. Label inputs as A, B, and B_{in}; label outputs Di and B_o.

12. Draw the truth table for the full subtractor. Label inputs A, B, and B_{in}; label outputs Di and B_o.

13. The 1s complement and end-around carry method of subtraction is used when ___?___ (adders, half subtractors, full subtractors) are used in a 4-bit subtractor.

14. Half adders and ___?___ (AND, OR, XNOR) gates can be used to form a 2-bit adder.

15. Full adders and ___?___ (AND, OR, INVERTER) gates can be used to form a 4-bit subtractor.

16. On a parallel subtractor, the B_o output of one full subtractor is connected to the ___?___ (A, B, B_{in}) input of the next subtractor.

9-8 4-BIT ADDER/SUBTRACTOR

Now that we know we can use full adders to add and subtract, let us design a system which will add and subtract. We shall start with the subtractor system in Fig. 9-14. To make this system a *4-bit adder*, we need only bypass temporarily the four inverters and disconnect the end-around carry line. The subtractor system has been redrawn in Fig. 9-15. Four XOR gates have replaced the inverters. When a 0 is placed in input A of the XOR gates, the bits from the problem pass through the XOR gate with no change (see the truth table at the lower left, Fig. 9-15). With the control at 0, the unit adds the binary numbers $A_3A_2A_1A_0$ to $B_3B_2B_1B_0$. The result (up to a sum of 1111) appears at the output indicators. The 0 on the control (add position) also disables the AND gate and does not permit the end-around carry.

To make the unit in Fig. 9-15 a *4-bit subtractor*, the control must be in the 1 position. This will cause the XOR gate to act as an inverter for the B inputs to the full adders. This complementing is seen in the truth table in the lower left, Fig. 9-15. The 1 at the control also activates the AND gate, so information from the 8s full adder can take the end-around carry line back to the 1s full adder. This subtractor will subtract binary input number $B_3B_2B_1B_0$ from $A_3A_2A_1A_0$. The difference will appear in binary form on the output indicators. Remember that we are using the 1s complement and end-around carry method of subtraction in this circuit. The XOR gates do the complementing, and the end-around carry line is labeled.

Considering that multiplication is just repeated addition and division is repeated subtraction, you can see how important the adder/subtractor circuit is in digital electronics.

Serial adder

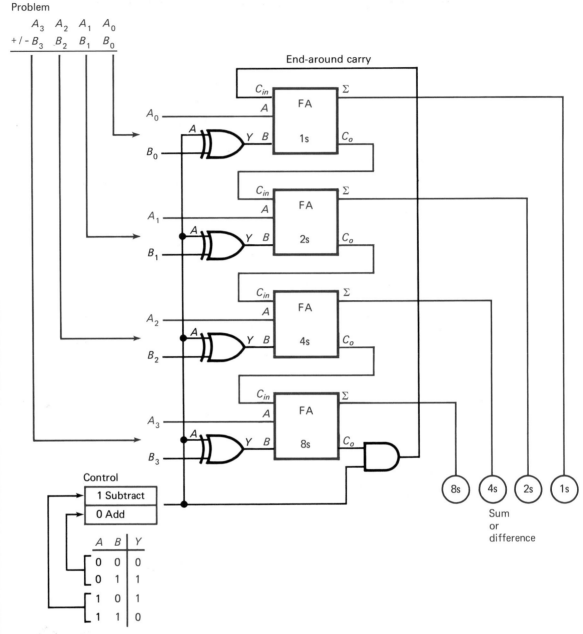

Problem

$$A_3 \quad A_2 \quad A_1 \quad A_0$$
$$+/- \quad B_3 \quad B_2 \quad B_1 \quad B_0$$

End-around carry

A	B	Y
0	0	0
0	1	1
1	0	1
1	1	0

Control

1 Subtract

0 Add

Sum or difference

Fig. 9-15   Combination adder/subtractor circuit.

9-9 SERIAL ADDER SYSTEM

Until now we have been discussing parallel adders. A parallel adder needs a full adder for each binary column added. Another method of adding uses a *serial adder*. In a serial adder only one full adder is required. When used with shift registers, a serial adder system might appear like the one in Fig. 9-16. At the top diagram are two shift registers (A and B) feeding the A and B inputs of a single full-adder circuit. The output sum will be accumulated in the sum register at the right. The top diagram shows the A and B registers loaded with the numbers to be added: binary $A_3A_2A_1A_0$ and $B_3B_2B_1B_0$. On the first clock pulse the 1s place (A_0 and B_0) digits are added, with the result appearing on the sum register (S_0). On the second clock pulse the 2s place digits are added with the carry from the delay flip-flop. The result (S_1) is shifted into the sum register. On the third clock pulse inputs A_2, B_2, and C_{in} are added, with the result appearing as

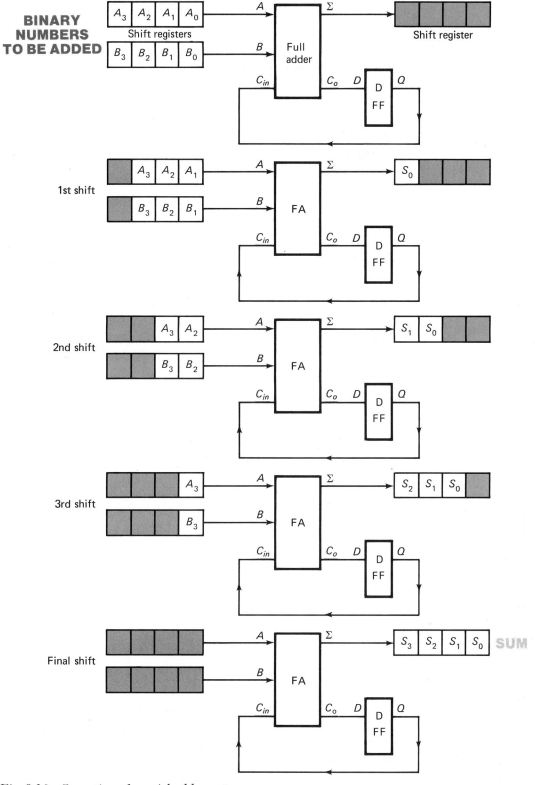

Fig. 9-16   Operation of a serial adder system.

S_2 on the sum register. The fourth and final clock pulse adds inputs A_3, B_3, and C_{in}. The result is shifted into the sum register, and the addition problem is complete. The sum $(S_3S_2S_1S_0)$ is the binary number in the right sum register after all four clock pulses.

Observe that the clock input of the three shift registers and the delay flip-flop in Fig. 9-16 were connected. Also note that only two binary bits were added at a time. The bits were fed to the full adder in a serial fashion from the shift registers. This is why this system is called a *serial adder*. You can thus appreciate why it is important that *only four* clock pulses be used to complete the addition. More than four pulses would leave bits in the sum register out of position, and the answer would be incorrect.

9-10 IC ADDERS

Integrated circuit manufacturers produce several adders. One useful arithmetic IC is the 7483 4-bit binary full adder. A block symbol for the 7483 *IC adder* is drawn in Fig. 9-17. The problem of adding the two 4-bit binary numbers ($A_3A_2A_1A_0$ and $B_3B_2B_1B_0$) is shown being entered into the eight inputs of the 7483

IC. Notice a difference in numbering systems on the problem and the IC. For adding just two 4-bit numbers, the C_o input (end-around carry input) is held at 0. The sum outputs are shown attached to output indicators. The C_4 output is attached to the 16s output indicator. This binary adder will indicate a sum as high as 11110 (decimal 30) when adding binary 1111 to 1111.

Internally the 7483 IC adder is organized very much like the unit in Fig. 9-14 (without the four inverters). The C_4 (carry out) on the 7483 IC is the same as the C_o on the 8s full adder in Fig. 9-14. The C_o (carry input) on the 7483 IC is the same as the C_{in} on the 1s full adder in Fig. 9-14.

The 7483 IC adder can be cascaded by connecting the C_4 output of the first IC to the C_o (carry input) of the next 7483 IC. With two 7483 IC adders cascaded, an 8-bit binary adder is produced. The 7483 IC can also be a 4-bit subtractor using the same arrangement shown in Fig. 9-14. The B inputs are inverted or complemented, and C_4 (carry output) is the end-around carry line to the C_o (carry input) of the 7483 IC. The 7483 IC may also be used in the adder/subtractor circuit diagramed in Fig. 9-15. Of course, extra ICs would be used for the XOR and AND gates.

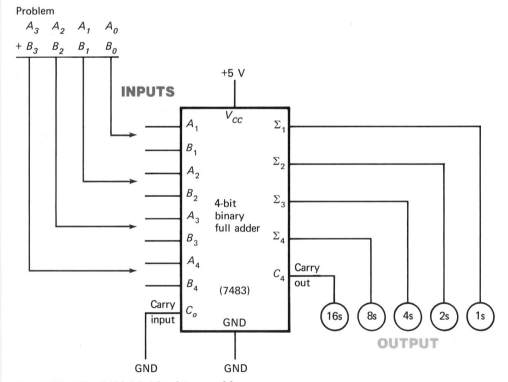

Fig. 9-17 The 7483 IC 4-bit binary adder.

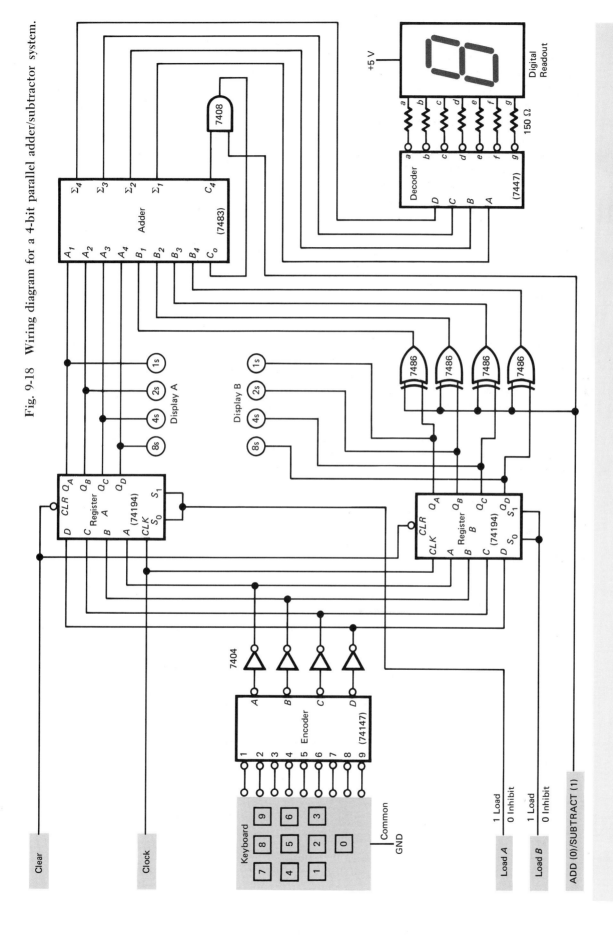

Fig. 9-18 Wiring diagram for a 4-bit parallel adder/subtractor system.

9-11 PARALLEL ADDER/SUBTRACTOR SYSTEM

The serial adder we studied in Sec. 9-9 used shift registers and an adder to form a system. In this section we shall diagram a digital electronics system with the parallel adder as its heart. Figure 9-18 is a wiring diagram of a *parallel adder/subtractor system*. This system employs parts you have already used. The inputs are shown at the left; the output is the digital readout (seven-segment display) at the right.

The following steps might be used to operate the parallel adder/subtractor system in Fig. 9-18. First, activate the *CLR* input control to clear both registers (A and B) to 0000. Second, set the add/subtract control to the proper mode (let us use 0 for Add). Next, load each register individually with the load controls. To load register A, place the load A control to 1 (load B to 0). Press a keyboard number while pulsing the *CLK* input once. The binary number you loaded in register A will appear on display A. To load register B, place the load B control to 1 (load A to 0). Press a keyboard number while pulsing the *CLK* input once. This second binary number should appear in register B as indicated on display B. The 7483 IC 4-bit adder works immediately, and the *sum* appears on the digital display.

The subtractor will subtract the contents of register B from the binary number in register A. The procedure for subtracting with the system in Fig. 9-18 is the same as for adding *except* that the add/subtract control is set in the subtract position(1). This activates the AND and XOR gates for the 1s complement and end-around carry subtraction. The digital display will give the *difference* between register A and register B.

The 74147 encoder changes the decimal input from the keyboard into binary numbers in Fig. 9-18. The 7404 IC inverts the outputs of the 74147 encoder. From the inverters the binary is fed into the parallel load data inputs of both registers A and B. With one clock pulse the parallel data at the input of the register is transferred and stored in the registers (if S_0 and S_1 are at 1). Displays A and B indicate what binary numbers are stored in registers A and B; the binary numbers stored in the registers are applied to the inputs of the 7483 4-bit

adder. The adder adds binary numbers $A_4A_3A_2A_1$ and $B_4B_3B_2B_1$, and the sum appears in binary at the outputs of the 7483 adder. The sum is decoded from binary to seven-segment code by the 7447 decoder. The decimal sum appears on the digital readout.

It is quite common to have an arithmetic unit serve as part of the *central processing unit* (often referred to as the CPU). A block diagram of this digital system is shown in Fig. 8-1. The processing unit in that diagram is your adder/subtractor.

9-12 BINARY MULTIPLICATION

In grade school you learned how to multiply. You learned to lay out your multiplication problem similar to that in Fig. 9-19(a). You learned that the top number was called the *multiplicand*, the bottom number the *multiplier*. The solution to the problem was called the *product*. The product of 7×4, then, is 28, as shown in Fig. 9-19(a).

Figure 9-19(b) shows that multiplication is really just *repeated addition*. The problem $7 \times 4 = 28$ is represented by the multiplicand (7) being added four times. The 4 is the amount of the multiplier. The product produced is 28.

If you were to multiply 54×14, the repeated addition system would be complicated and take a long time. The multiplicand (54) would have to be added 14 times to get a product of 756. Most of us were taught to multiply 54×14 in the manner shown in Fig. 9-20(a). To solve the multiplication problem

$$\begin{array}{r} 7 \\ \times\,4 \\ \hline 28 \end{array}$$ Multiplicand Multiplier Product

(a)

$$7 + 7 + 7 + 7 = 28$$

Multiplicand — Product — Multiplier = 4

(b)

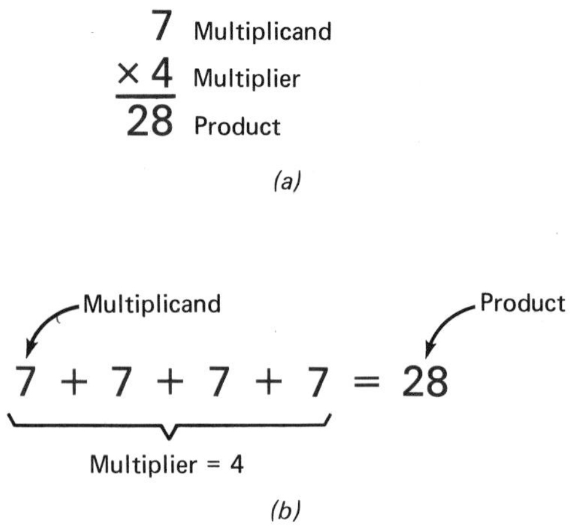

Fig. 9-19 (*a*) Decimal multiplication problem. (*b*) **Multiplying using the repeated addition method.**

54×14, we first multiply the multiplicand 54 by 4. This resulted in the first partial product (216) shown in Fig. 9-20(b). Next we multiply the multiplicand 54 by 1. Actually the multiplicand (54) is multiplied by a multiplier of 10, as shown in Fig. 9-20(c). The second partial product is 540. The first and second partial products (216 and 540) are then added for a final product of 756. It is normal to omit the 0 in the second partial product, as in Fig. 9-20(a).

It is important to notice the *process* in the problem in Fig. 9-20. The multiplicand is first multiplied by the LSD of the multiplier. This gives the first partial product. The second partial product is then calculated by multiplying the multiplicand by the MSD of the multiplier. The two partial products are then added, producing the final product. This same process is used in *binary multiplication*.

Binary multiplication is much simpler than decimal multiplication. The binary system has only two digits (0 and 1), which makes the rules for multiplying simple. Figure 9-21(a) shows the rules for binary multiplication.

Multiplication with binary numbers is done just as with decimal numbers. Figure 9-21(b) details a problem where binary 111 (say "one

Rules for binary multiplication

$$\begin{array}{cccc} 0 & 0 & 1 & 1 \\ \times 0 & \times 1 & \times 0 & \times 1 \\ \hline 0 & 0 & 0 & 1 \end{array}$$

(a)

Decimal	Binary	
7	111	Multiplicand
×5	×101	Multiplier
35	111	First partial product
	000	Second partial product
	111	Third partial product
	100011	Product

(b)

Fig. 9-21 (a) **Rules for binary multiplication.** (b) **Sample multiplication problem.**

one one") is multiplied by binary 101 (say "one zero one"). First, the multiplicand (111) is multiplied by the 1s bit of the multiplier. The result is the first partial product shown as 111 in Fig. 9-21(b). Next, the multiplicand (111) is multiplied by the 2s bit of the multiplier. The result is the second partial product (0000). Notice that the least significant bit (LSB) of the second partial product 0000 is left off in Fig. 9-21(b). Third, the multiplicand (111) is multiplied by the 4s bit of the multiplier. The result is the third partial product of 11100, shown in Fig. 9-21(b) as 111, with the two blank spaces filling the 1s and 2s places. Finally, the first, second, and third partial products are added, resulting in a product of binary 100011. Notice that the same multiplication problem in decimal is shown at the left of Fig. 9-21(b) for your convenience. The binary product 100011 equals the decimal product 35.

An example of another binary multiplication problem is shown in Fig. 9-22. At the left the problem is in the familiar decimal form; the same problem is repeated in binary form at the right, where here binary 11011 is multiplied by 1100. As in decimal multiplication, the zeros in the multiplier can simply be brought down to hold the 1s and 2s places in the binary number. The binary product is shown as 101000100, which equals decimal number 324.

$$\begin{array}{r} 54 \\ \times 14 \\ \hline 216 \\ 54 \\ \hline 756 \end{array}$$

Multiplicand
Multiplier

Product

(a)

$$\begin{array}{r} 54 \\ \times 14 \\ \hline 216 \end{array}$$

First partial product

(b)

$$\begin{array}{r} 54 \\ \times 10 \\ \hline 216 \\ 540 \end{array}$$

First partial product
Second partial product

(c)

Fig. 9-20 (a) **Decimal multiplication problem.** (b) **Calculating the first partial product.** (c) **Calculating the second partial product.**

Decimal	Binary	
27	1 1 0 1 1	Multiplicand
× 12	× 1 1 0 0	Multiplier
54	1 1 0 1 1 0 0	Third partial product
27	1 1 0 1 1	Fourth partial product
3 2 4	1 0 1 0 0 0 1 0 0	Product

Fig. 9-22   Sample multiplication problem.

You will gain some experience in solving binary multiplication problems by answering the questions at the end of the chapter.

9-13 BINARY MULTIPLIERS

We can multiply numbers by repeated addition, as illustrated by 7 multiplied by 4 in Sec. 9-12. The multiplicand (7) could be added four times to obtain the product of 28. A block diagram of a circuit that will perform repeated addition is shown in Fig. 9-23. The multiplicand is held in the top register. In our example the multiplicand is a decimal 7 or a binary 111 (say "one one one"). The multiplier is held in the down counter shown on the left in Fig. 9-23. The multiplier in our example is a decimal 4 or a binary 100. The lower product register will hold the product.

The repeated addition technique is shown in operation in Fig. 9-24. This chart shows how the multiplicand (binary 111) is multiplied by the multiplier (binary 100). The product register is cleared to 00000. After one count downward, the partial product of 00111 (decimal 7) is in the product register. After the second count downward, a partial product of 01110 (decimal 14) appears in the product register. After the third count downward, a partial product of 10101 (decimal 21) appears in the product register. After the fourth downward count, the *final product* of 11100 (decimal 28) appears in the product register. The multiplication problem (7 × 4 = 28) is complete. The circuit in Fig. 9-23 has added 7 four times for a total of 28.

This type of circuit is not widely used because of the long time it takes to do the repeated addition when large numbers are multiplied. A more practical method of multiplying is now detailed.

A common method of multiplying using digital electronic circuits is the "add and shift"

method. Figure 9-25 shows a binary multiplication problem. In this problem binary 111 is multiplied by 101 (7 × 5 in decimal). This hand-done procedure is standard, except for the temporary product in line 5. Line 5 was added to help you understand how multiplication might be done by digital circuits. Close observation of binary multiplication shows the following three important facts:

1. Partial products are always 000 if the multiplier is 0 or equal to the multiplicand if the multiplier is 1.
2. Product register needs twice as many bits as the multiplicand.
3. The first partial product is shifted one place to the right compared to the second partial product when adding.

You can observe each characteristic by looking at the sample problem in Fig. 9-25.

The important characteristics of longhand multiplication were just given. A binary multiplication circuit can be designed by using

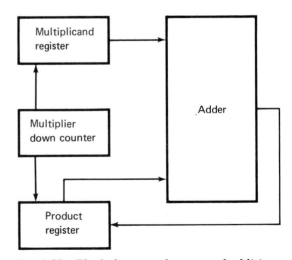

Fig. 9-23   Block diagram of a repeated addition-type multiplier system.

	Load with binary	After 1 down count	After 2 down counts	After 3 down counts	After 4 down counts
Multiplicand register	111	111	111	111	111
Multiplier counter	100	011	010	001	000
Product register	00000	00111	01110	10101	11100
	Load				Stop

Fig. 9-24 Multiplying binary 111 and 100 using the repeated addition circuit.

Binary multiplication

these characteristics. Figure 9-26(a) shows a circuit that will do binary multiplication. Notice that the multiplicand (111) is loaded into the register at the upper left. The accumulator register is cleared to 0000. The multiplier (101) is loaded into the register at the lower right. Notice too that the accumulator and the multiplier are considered together. This is shown by the shading connecting the two registers.

Let us use the circuit in Fig. 9-26(a) to go through the detailed procedure for multiplying. The diagram in Fig. 9-26(b) is a step-by-step review of how binary 111 is multiplied by 101 using the add and shift method. The binary 111 is loaded into the multiplicand register. The accumulator and multiplier registers are loaded in step A in Fig. 9-26(b). Step B shows the 0000 and the 111 from the accumulator and multiplicand registers being added due to the 1 being applied to the control line. This is comparable to line 3 of the multiplication problem in Fig. 9-25. Step C shifts both the accumulator and multiplier registers to the right one place. The LSB of the multiplier (1) is shifted out the right end and lost.

Step D represents another *add* step. This time a 0 is applied to the control line. A 0 on the control line means *no* addition. The register contents remain the same. Step D is comparable to lines 4 and 5, Fig. 9-25. Step E shows the registers being shifted one place to the right. This time the 2s bit of the multiplier is lost as it is shifted out the right end of the register. Step F shows the 4s bit of the multiplier (1) commanding the adder to add. The accumulator contents (0001) and the multiplicand (111) are added. The result of that addition is deposited into the accumulator register (1000). This step is comparable to the left section of lines 5 to 7, Fig. 9-25. Step G is the final step in the add and shift multiplication; it shows a single shift to the right for both registers. The MSB (4s bit) of the multiplier is lost out of the right end of the register. The final product appears across both registers as 100011. Binary 111 multiplied by 101 resulted in a product of 100011 ($7 \times 5 = 35$ in decimal). This final product produced by the multiplier circuit is the same result we got in line 7, Fig. 9-25, when we multiplied by hand.

Line 1	1 1 1	Multiplicand
Line 2	× 1 0 1	Multiplier
Line 3	1 1 1	First partial product
Line 4	0 0 0	Second partial product
Line 5	0 1 1 1	Temporary product (line 3 + line 4)
Line 6	1 1 1	Third partial product
Line 7	1 0 0 0 1 1	Product

Fig. 9-25 Binary multiplication problem.

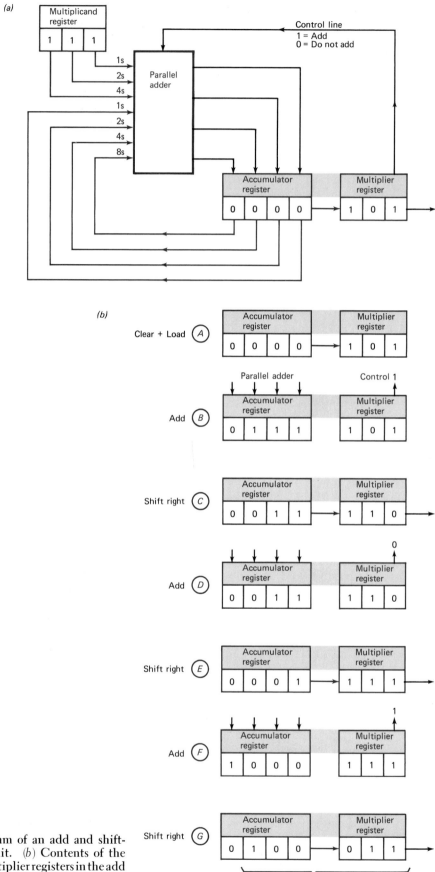

(a)

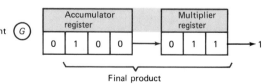

Fig. 9-26 (a) Diagram of an add and shift-type multiplier circuit. (b) Contents of the accumulator and multiplier registers in the add and shift multiplier circuit.

Two types of multiplier circuits are illustrated. The first used repeated addition to arrive at the product. That system was shown in Fig. 9-23. The second circuit used the add and shift method of multiplying. The add and shift system was shown in Fig. 9-26.

In many computers *the procedure* such as the add and shift method can be programmed into the machine. Instead of permanently wiring the circuit, we simply *program* or instruct the computer to follow the procedure shown in Fig. 9-26(*b*). We are thus using *software* (a program) to do multiplication. This use of software cuts down on the amount of electronic circuits needed in the CPU of a computer.

Summary

1. Arithmetic circuits such as adders and subtractors are combinational logic circuits constructed with logic gates.
2. The basic addition circuit is called a half adder. Two half adders and an OR gate make a full adder.
3. The basic subtraction circuit is called a half subtractor. Two half subtractors and an OR gate make a full subtractor.
4. Adders (or subtractors) can be wired together or used alone to form parallel or serial adders.
5. A 4-bit parallel adder will add two 4-bit binary numbers at one time. This adder would contain a single half adder (1s place) and three full adders.

6. By using the 1s complement and end-around carry method of subtraction, an adder can be used for binary subtraction.
7. By using AND and XOR gates, the adder and subtractor functions can be combined in one device.
8. Manufacturers produce several arithmetic ICs, such as the 7483 4-bit binary adder.
9. Adder/subtractor units are often part of the CPU of calculating machines.
10. Binary multiplication performed by digital circuits may use repeated additions or the add and shift method.

Questions

1. Do binary addition problems *a* to *h* (*show your work*):

    a. 101 + 011 = ____?____   *e*. 1000 + 1000 = ____?____

    b. 110 + 101 = ____?____   *f*. 1001 + 0111 = ____?____

    c. 111 + 111 = ____?____   *g*. 1010 + 0101 = ____?____

    d. 1000 + 0011 = ____?____   *h*. 1100 + 0101 = ____?____

2. Draw a block diagram for a half adder (label two inputs and two outputs).

3. Draw a block diagram for a full adder (label three inputs and two outputs).

4. Do binary subtraction problems *a* to *h* (show your work):

    a. 1100 − 0010 = ____?____   *e*. 10000 − 0011 = ____?____

    b. 1101 − 1010 = ____?____   *f*. 1000 − 0101 = ____?____

    c. 1110 − 0011 = ____?____   *g*. 10010 − 1011 = ____?____

    d. 1111 − 0110 = ____?____   *h*. 1001 − 0010 = ____?____

5. Draw a block diagram for a half subtractor (label two inputs and two outputs).

6. Draw a block diagram for a full subtractor (label three inputs and two outputs).

7. Draw a block diagram of a 2-bit parallel adder (use a half and a full adder).

8. Draw a block diagram of a 3-bit parallel subtractor (use three full adders and three inverters).

9. Draw a block diagram of a 3-bit parallel adder/subtractor (use three full adders, three XOR gates, and one AND gate).

10. Do binary subtraction problems a to h using the 1s complement and end-around carry method (show your work):
    a. $111 - 101 =$ ___?___   e. $1011 - 1010 =$ ___?___
    b. $1000 - 0011 =$ ___?___   f. $1100 - 0110 =$ ___?___
    c. $1001 - 0010 =$ ___?___   g. $1110 - 0100 =$ ___?___
    d. $1010 - 0100 =$ ___?___   h. $1111 - 0111 =$ ___?___

11. Draw a logic symbol diagram of a 2-bit parallel adder using AND, OR, and XOR gates.

12. Draw a logic symbol diagram of a 2-bit parallel subtractor using AND, OR, and XOR gates and inverters.

13. Do binary multiplication problems a to h (show your work). Check your answers using decimal multiplication.
    a. $101 \times 011 =$ ___?___   e. $1010 \times 011 =$ ___?___
    b. $111 \times 011 =$ ___?___   f. $110 \times 111 =$ ___?___
    c. $1000 \times 101 =$ ___?___   g. $1100 \times 1000 =$ ___?___
    d. $1001 \times 010 =$ ___?___   h. $1010 \times 1001 =$ ___?___

14. List two methods of doing binary multiplication with digital electronic circuits.

15. If the CPU of your computer has only an adder and shift registers, how can you still do binary multiplication with this machine?

Answers to Self Tests

1. a. 1110   f. 0101
   b. 1011   g. 0100
   c. 10001 h. 11000
   d. 10011 i. 11010
   e. 0111   j. 11000

2.
   A ─┐ HA ├─ Sum
   B ─┘   ├─ Carry Out

3.

A	B	Σ	C_o
0	0	0	0
0	1	1	0
1	0	1	0
1	1	0	1

4.
   A, B → XOR → Σ
        → AND → C_o

5.
   C_{in}, A, B → FA ├─ Sum
                       ├─ Carry Out

6.

C_{in}	A	B	Σ	C_o
0	0	0	0	0
0	0	1	1	0
0	1	0	1	0
0	1	1	0	1
1	0	0	1	0
1	0	1	0	1
1	1	0	0	1
1	1	1	1	1

7. a. 0001   f. 0101
   b. 0010   g. 1111
   c. 0000   h. 0011
   d. 1011   i. 0111
   e. 0001   j. 0111

8. a. 0001   f. 0101
   b. 0101   g. 0010
   c. 0001   h. 0111
   d. 0010   i. 0110
   e. 0010   j. 1011

9.
   A ─┐ HS ├─ Difference
   B ─┘   ├─ Borrow Out

10.

A	B	Di	B_o
0	0	0	0
0	1	1	1
1	0	1	0
1	1	0	0

11.
   B_{in} ─┐   ├─ Difference
   A ───── FS
   B ───────┘   ├─ Borrow Out

12.

A	B	B_{in}	Di	B_o
0	0	0	0	0
0	0	1	1	1
0	1	0	1	1
0	1	1	0	1
1	0	0	1	0
1	0	1	0	0
1	1	0	0	0
1	1	1	1	1

13. Adders
14. OR
15. inverter
16. B_{in}

Memories

- The flip-flop forms the basic "memory cell" in many semiconductor memories. You have already used a shift register (several flip-flops wired together) to form a temporary memory. The types of semiconductor memories we shall investigate in this chapter are the RAM, ROM, and PROM.

Many pocket-sized calculators contain an electronic *memory*. A given number can be stored in this electronic memory by the simple press of the "store" button on the keyboard. The number can be recalled from memory by another press of the "recall" key. Human memory works the same way. You store information by learning, and you recall information by remembering. Most digital electronic systems use many types of memories. Another term for memory is *storage*.

A complicated digital system such as a computer will contain internal storage or memories. These probably will be in the form of *magnetic core* and *semiconductor* (IC) memories. Outside the computer, information will be stored on *bulk-storage* memory devices, including punched cards, punched paper tape, magnetic tape, and disc and drum storage. Except in computers you will probably run into semiconductor memories most often.

10-1 RANDOM-ACCESS MEMORY (RAM)

One type of semiconductor memory device used in digital electronics is the *random-access memory* (RAM). The RAM is a memory that you can "teach." After the "teaching-learning" process (called *writing*), the RAM remembers the information for a while, the RAM's stored information can be recalled or "remembered" at any time. We say that we can *write* information (0s and 1s) into the memory or *read out* or recall information.

The random-access type of memory is also called a *read/write memory* or *scratch pad memory*.

A semiconductor memory with 64 positions in which to place 0s and 1s is illustrated in Fig. 10-1. The 64 squares (mostly blank) at the right represent the 64 positions that can be filled with data. Notice that the 64 bits are organized into 16 groups called *words*. Each word contains 4 *bits* of information. This memory is said to be organized as a 16×4 memory. That is, it contains 16 words, and each word is 4 bits long. A 64-bit memory could be organized as a 32×2 memory (32 words of 2 bits each), a 64×1 memory (64 words of 1 bit each), or an 8×8 memory (8 words of 8 bits each).

The memory in Fig. 10-1 looks very much like a truth table on a scratch pad. On the table after Word 3 we have written the contents of Word 3 (0110). We say we have stored or *written* a word into the memory; this is the write operation. To see what is in the memory at word location 3, just read from the table in Fig. 10-1; this is the read operation. The write operation is the process of putting new information into the memory. The read operation is the process of taking information out of the memory. The read operation is also referred to as the *sense* operation because it senses or reads the contents of the memory.

You could write any combination of 0s and 1s in the table in Fig. 10-1, rather like writing on a scratch pad. You could then read any word(s) from the memory, as from a scratch pad. Notice that the information in the

Address	Bit D	Bit C	Bit B	Bit A
Word 0				
Word 1				
Word 2				
Word 3	0	1	1	0
Word 4				
Word 5				
Word 6				
Word 7				
Word 8				
Word 9				
Word 10				
Word 11				
Word 12				
Word 13				
Word 14				
Word 15				

Fig. 10-1 Organization of a 64-bit memory.

memory remains even after it is read. Now it should be obvious why this memory is sometimes called a 64-bit scratch pad memory. The memory has a place for 64 bits of information, and the memory can be written into or read from very much like from a scratch pad.

The memory in Fig. 10-1 is called a random-access memory (RAM) because you can go directly to Word 3 or Word 15 and read its contents. In other words, you have access to any bit (or word) at any instant. You merely skip down to its word location and read that word. A location in the memory such as Word 3 is referred to as the storage location or *address*. In the case of Fig. 10-1, the address of Word 3 refers to 4 bits of information (bit *D*, bit *C*, bit *B*, and bit *A*).

The RAM cannot be used for permanent memory because it loses its data when the power to the IC is turned off. The RAM is considered a *volatile* memory because of this loss of data. Volatile memories are thus used for the *temporary* storage of data. However, some memories are permanent; they do not "forget" or lose their data when the power goes off. Such permanent memories are called *nonvolatile* storage devices.

RAMs are used where only a temporary memory is needed. RAMs are used for calculator memories, buffer memories, or cache memories in computers.

10-2 AN IC RAM

The 7489 *read/write random-access memory* is a 64-bit storage unit in IC form. Figure 10-2 is a diagram of the 7489 RAM. Remember that inside this RAM the memory cells are arranged as in the table in Fig. 10-1. The memory can hold 16 words; each word in the 7489 is 4 bits long. The 7489 RAM is organized as a 16 × 4-bit RAM.

Let us first *write* data into the 7489 memory. Suppose we want to write a 0110 into the Word 3 location, as shown in Fig. 10-1. The address for Word 3 is $D = 0$, $C = 0$, $B = 1$, and $A = 1$. Locate Word 3 in the memory by placing a binary 0011 (decimal 3) on the *address inputs* of the 7489 IC (see Fig. 10-2). Next, place the correct input data at the *data inputs*. To enter 0110, place a 0 at input A, a 1 at input B, a 1 at input C, and a 0 at input D. Next, place the *write enable* at 0. Last, place the *memory enable* at 0; the data is written into the memory in the storage location called Word 3. This was the write process.

Now let us *read* or sense what is in the memory. If we want to read out the data at Word 3, we set the address inputs to binary 0011 (decimal 3). The write enable input should be in the read position or at 1. The memory enable should be at logical 0. The *data outputs* will indicate 1001. This output is the complement of the actual memory content, which is 0110. Inverters could be attached to the outputs of the 7489 IC to make the output data the same as that in the memory. This was the read process.

To write or read at a memory location other than Word 3, we change the binary input on the address input. To get to Word 0 we apply binary 0000 to the address input in Fig. 10-2. For Word 9 we apply 1001 to the address input.

You will find that although different manufacturers use various labels for the inputs and outputs on this IC, all 7489 ICs have the inputs and outputs shown in Fig. 10-2.

The 7489 RAM is just one example of a semiconductor memory in IC form. Many other semiconductor memories are available from IC manufacturers. Semiconductor memories are relatively new, but they are being widely used in many applications. They have the advantage of being inexpensive, small, reliable, and can operate at high speeds.

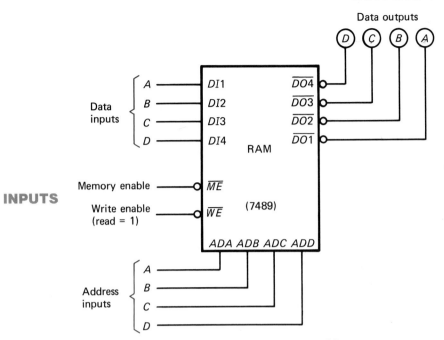

Program the memory

Fig. 10-2   Logic diagram of the 7489 random-access memory IC.

10-3 USING A RAM

We need some practice in using the 7489 read/write RAM. Let us *program* it with some usable information. To program the memory is to write in the information we want in each memory cell.

Probably you cannot remember how to count from 0 to 15 in the Gray code, so let us take the Gray code and program it into the 7489 RAM. The RAM will remember the Gray code for us, and we can then use the RAM to convert from binary numbers to the Gray-code numbers.

Table 10-1 shows the Gray coded numbers from 0 to 15. For convenience binary numbers are also included in Table 10-1. The 64 ones and zeros in the Gray code number column of the table must be written into the 64-bit RAM. The 7489 IC is perfect for this job because it contains 16 words; each word is 4 bits long. This is the same pattern we have in the Gray code column of Table 10-1. The decimal number in the table will be the Word number (see Fig. 10-1). The binary number column is the number applied to the address input of the 7489 RAM (see Fig. 10-2). The Gray code number is applied to the data inputs of the RAM (see Fig. 10-2). When the memory enable and write enable inputs are activated, the Gray code will be written into

the 7489 RAM. The RAM will remember this code as long as the power is not turned off.

After the 7489 RAM is programmed with the Gray code, it is a code converter. Figure 10-3(a) shows the basic system. Notice that we shall input a binary number. The code converter will read out the equivalent Gray

Table 10-1   Gray code.

Decimal number	Binary number	Gray code number
0	0000	0000
1	0001	0001
2	0010	0011
3	0011	0010
4	0100	0110
5	0101	0111
6	0110	0101
7	0111	0100
8	1000	1100
9	1001	1101
10	1010	1111
11	1011	1110
12	1100	1010
13	1101	1011
14	1110	1001
15	1111	1000

Volatile
memory

Nonvolatile
memory

Read-only
memory (ROM)

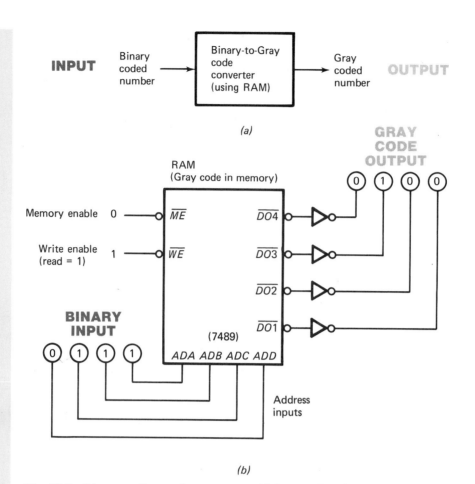

Fig. 10-3 Binary-to-Gray code converter. (a) System diagram. (b) Wiring diagram using RAM.

code number. The system is a binary-to-Gray code converter.

How do you convert binary 0111 (decimal 7) to the Gray code? Figure 10-3(b) shows the binary number 0111 being applied to the address inputs of the 7489 RAM. The memory enable input is at 0. The write enable input is in the read position (logical 1). The 7489 IC then reads out the stored Word 7 in inverted form. The four inverters complement the output of the RAM. The result is the correct Gray code output. The Gray code output for binary 0111 is shown as 0100 in Fig. 10-3(b). You may input any binary number from 0000 to 1111 and get the correct Gray code output.

The binary-to-Gray code converter in Fig. 10-3 works fine. It demonstrates how you can program and use the 7489 RAM. But it is not very practical because the RAM is a volatile memory. If the power is turned off for even an instant, the storage unit loses all its memory and "forgets" the Gray code. We say the memory has been *erased*. You would then have to again program or teach the Gray code to the 7489 RAM.

10-4 READ-ONLY MEMORY (ROM)

A disadvantage of the RAM is that its memory is volatile. When the power is turned off the RAM loses all its data. Some semiconductor memories have *nonvolatile* memories. A nonvolatile memory is a permanent memory that never forgets its data. One type of non-volatile semiconductor memory is called the *read-only memory* (ROM). The ROM has a pattern of 0s and 1s programmed into the IC permanently by the manufacturer. If the program is permanently in a ROM, then it is not possible to write new information into the memory; hence the title of read-only for this storage device.

ROMs may be organized like RAMs. For instance, a 256-bit ROM might be organized as a 32 × 8 storage unit (32 words that are each 8 bits long). A 1024-bit ROM might be organized as a 256 × 4 storage unit (256 words that are each 4 bits long).

A ROM can be used when the task always demands the same outputs for given inputs. The job could be done with a combinational logic circuit using gates. Using a ROM is sometimes easier and less expensive then using gates. ROMs usually do not use flip-flops as memory cells. Special circuits that can be set by the manufacturer at a logical 0 or 1 form the memory cells in ROMs.

10-5 USING A ROM

Suppose you have to design a device that will give the decimal counting sequence shown in Table 10-2: 1, 117, 22, 6, 114, 44, 140, 17, 0, 14, 162, 146, 134, 64, 160, 177, and then back to 1. These numbers are to read out on seven-segment displays and must appear in the order shown.

Knowing you will use digital circuits, you convert the decimal numbers to binary-coded decimal (BCD) numbers. This is shown in Table 10-2. You find you have 16 rows and 7 columns of logical 0s and 1s. This section forms a truth table. As you look at the truth table, the problem seems quite complicated to solve with logic gates or data selectors. You

decide to try a ROM. You think of the inside of a memory as a truth table. The BCD section of Table 10-2 reminds you that a memory organized as a 16 × 7 storage unit will do the job. This 16 × 7 ROM will have 16 words for the 16 rows on the truth table. Each word will contain 7 bits of data for the seven columns on the truth table. This would take a 112-bit ROM.

A 112-bit ROM is shown in Fig. 10-4. Notice it has four address inputs to select one of the 16 possible words stored in the ROM. The 16 different addresses are shown in the left columns of Table 10-3. Suppose the address inputs were binary 0000. Then the first line in Table 10-3 would show that the stored word would be 0 000 001 (*a* to *g*). After decoding in Fig. 10-4, this stored word would read out on the digital displays as a decimal 1 (100s = 0, 10s = 0, 1s = 1).

Let us consider another example. Apply binary 0001 to the address inputs of the ROM in Fig. 10-4. The second row on Table 10-3 shows us the stored word is 1 001 111 (*a* to *g*). When decoded this word would read out on the digital display as a decimal 117 (100s = 1, 10s = 1, 1s = 7). Remember that the 0s and 1s in the center section of Table 10-3 are *permanently* stored in the ROM. When the address at the left appears at the address input of the ROM, a row of 0s and 1s or word appears at the outputs.

You have solved the difficult counting sequence problem. Figure 10-4 diagrams the basic system to be used. The information in Table 10-3 shows the addressing and programming of the 112-bit ROM and the decoded BCD as a decimal readout. You would give the information in Table 10-3 to a manufacturer, who would custom make as many ROMs as you need with the correct pattern of 0s and 1s.

It is quite expensive to have just a few ROMs custom programmed by a manufacturer. You probably would not use the ROM if you did not have need for many of these memory units. Remember that this problem also could have been solved by a combinational logic circuit using logic gates.

Semiconductor memories usually come in 64-, 256-, 1024-, and 4096-bit units. A 112-bit memory is an unusual size. The 112-bit memory was used in the example because its truth table in Table 10-3 is exactly the truth table of the 7447 IC. You used the 7447 IC as

Table 10-2 Counting sequence problem.

Decimal readout			Binary coded decimal number		
100s	10s	1s	100s	10s	1s
		1	0	000	001
1	1	7	1	001	111
	2	2	0	010	010
		6	0	000	110
1	1	4	1	001	100
	4	4	0	100	100
1	4	0	1	100	000
	1	7	0	001	111
		0	0	000	000
	1	4	0	001	100
1	6	2	1	110	010
1	4	6	1	100	110
1	3	4	1	011	100
	6	4	0	110	100
1	6	0	1	110	000
1	7	7	1	111	111

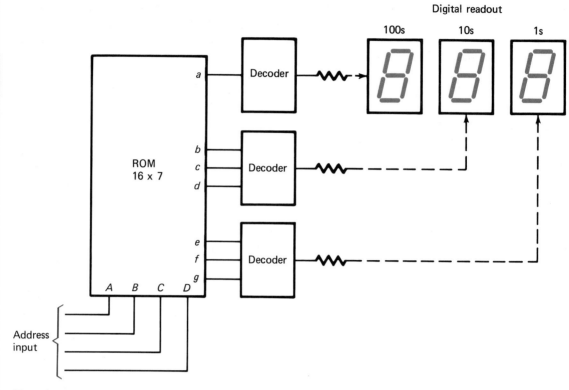

Fig. 10-4 System diagram for the counting sequence problem using a ROM.

Table 10-3 Counting sequence problem.

INPUTS				ROM OUTPUTS							Decimal readout		
Address or word location				100s	10s			1s					
				1s	4s	2s	1s	4s	2s	1s			
D	*C*	*B*	*A*	*a*	*b*	*c*	*d*	*e*	*f*	*g*	100s	10s	1s
0	0	0	0	0	0	0	0	0	0	1			1
0	0	0	1	1	0	0	1	1	1	1	1	1	7
0	0	1	0	0	0	1	0	0	1	0		2	2
0	0	1	1	0	0	0	0	1	1	0			6
0	1	0	0	1	0	0	1	1	0	0	1	1	4
0	1	0	1	0	1	0	0	1	0	0		4	4
0	1	1	0	1	1	0	0	0	0	0	1	4	0
0	1	1	1	0	0	0	1	1	1	1		1	7
1	0	0	0	0	0	0	0	0	0	0			0
1	0	0	1	0	0	0	1	1	0	0		1	4
1	0	1	0	1	1	1	0	0	1	0	1	6	2
1	0	1	1	1	1	0	0	1	1	0	1	4	6
1	1	0	0	1	0	1	1	1	0	0	1	3	4
1	1	0	1	0	1	1	0	1	0	0		6	4
1	1	1	0	1	1	1	0	0	0	0	1	6	0
1	1	1	1	1	1	1	1	1	1	1	1	7	7

a BCD-to-seven-segment decoder in Chap. 8. You will want to use the 7447 IC as a ROM in the lab.

Read-only memories are used for encoders, code converters, look-up tables, microprograms, and character generators.

10-6 PROGRAMMABLE READ-ONLY MEMORY (PROM)

We mentioned that a ROM is programmed by the manufacturer. A custom-made ROM is quite expensive. If you need only a few read-only memories, you can use a field *programmable read-only memory* (PROM). A PROM can be permanently programmed in your own shop by following the manufacturer's procedure to write the proper information into the PROM. The "one time" write process is done by selectively "blowing fuses" within the IC. Once programmed, the PROM acts exactly like a ROM. The PROM now contains a permanent pattern of 0s and 1s in its memory.

PROMs are organized in a manner similar to ROMs and RAMs. A 256-bit PROM may be a 32 × 8 storage unit (32 words each 8 bits long). A 1024-bit PROM may be a 256 × 4 unit (256 words each 4 bits long). Word lengths of 4 and 8 bits are very common in PROMs and ROMs.

A variation of the PROM is a *read-mostly memory*. The read-mostly memory is also called a reprogrammable read-only memory. These storage units are written into as on the PROM. If they need to be erased, special procedures are used. Some read-mostly memories may be erased by exposing the IC to ultraviolet light through a special window on the unit. Other read-mostly memories are erased electrically and use special MNOS transistors (metal nitride oxide silicon transistors). ROMs, PROMs, and read-mostly memories all are nonvolatile memories.

10-7 MAGNETIC CORE MEMORY

Digital computers need large amounts of internal and external data storage. Inside the computer semiconductor and *magnetic core memories* are used. The magnetic core memory is a time-proven, common method of storing data in the central memory of a computer.

The magnetic core memory is based upon the characteristics of the tiny *ferrite core*. A

ferrite core is a small doughnut-shaped piece of ferromagnetic material, such as iron. The core is baked and compressed into a ceramic-like doughnut. Figure 10-5(a) shows a highly magnified view of a ferrite core. A typical core might measure about $\frac{1}{16}$ in. across.

The ferrite core is used as a small magnet. Figure 10-5(b) shows a *write wire* threaded through the ferrite core. When current passes through the write wire in a given direction, the magnetic flux is shown traveling in the counterclockwise (ccw) direction. The magnetic flux direction is shown by an arrow on the core. We have defined the situation in Fig. 10-5(b) as a logical 1. In other words, counterclockwise movement of the magnetic flux in the core means it contains a logical 1.

Figure 10-5(c) shows the current being reversed. With the $-I$ pulse we find that the magnetic flux in the core reverses. The magnetic flux is now traveling in a clockwise (cw) direction in the core. The situation in Fig. 10-5(c) is defined as a logical 0; that is, a logical 0 is stored in the core when the flux travels in a clockwise direction. With no current flow in the write wire the ferrite core still is a magnet. Depending upon which direction the core is magnetized, it remembers a logical 0 or 1. Figure 10-5(d) shows the ferrite core with

Programmable read-only memory (PROM)

Read-mostly memory

Magnetic core memory

Ferrite cores

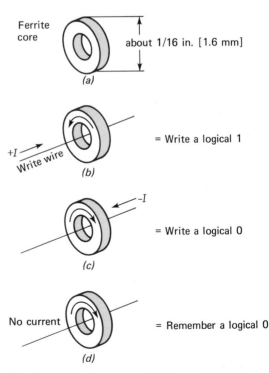

Fig. 10-5 Ferrite core. (*a*) **Size.** (*b*) **Writing a 1.** (*c*) **Writing a 0.** (*d*) **Remembering a 0.**

**Coincident-
current memory**

no current flowing in the write wire. The core still has magnetic flux moving in a clockwise direction. We say the core is storing a logical 0.

In Fig. 10-5 we saw how we write data into a magnetic core and how it remembers by staying magnetized. Like a flip-flop circuit, the ferrite core will always be either a logical 0 or 1. You have seen that the magnetic core memory is a nonvolatile type memory. It will remember even when the power is turned off.

The *read* process requires another wire. The added wire is called a *sense* wire, as shown in Fig. 10-6(*a*). To *read* the contents of the ferrite core we apply a $-I$ pulse to the core, as shown in Fig. 10-6(*b*). Assuming the core is a logical 0, there would be *no change* in magnetic flux in the core. With no change in magnetic flux we would have no current or zero amperes *induced* in the sense wire. We would read that the core contained a 0.

In Fig. 10-6(*c*) we assume that the core contains a logical 1. This is shown with the counterclockwise arrow on the core. To read the contents of the core we apply a $-I$ pulse in Fig. 10-6(*d*). The magnetic flux changes direction (from ccw to cw), as shown by the arrow. When the magnetic flux *changes di-*

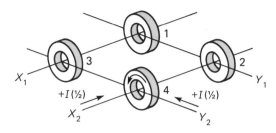

Fig. 10-7   Writing into a coincident-current core memory.

rection, a pulse is *induced* in the sense wire. The pulse in the sense wire tells us a logical 1 was stored in the core. Notice that the 1 which was stored in the ferrite core was destroyed by the read process. The core must be restored to the 1 state or the data will be lost.

When more than one ferrite core is used we need more than one write line. Figure 10-7 shows four ferrite cores, with two write wires passing through each. If we want to write a 1 in core 4, we pass one-half the amount of current needed through lines Y_2 and X_2. The two half currents add and write a 1 into core 4. Core 1 in Fig. 10-7 receives no current. Cores 2 and 3 receive only half current, and therefore nothing is written into these memory cells. A memory wired like the one in Fig. 10-7 is called a *coincident-current memory*. If we wanted to write a 1 into core 2, we would put $+\frac{1}{2}I$ in lines Y_1 and X_2. A $-I$ pulse through all the lines would reset all the cores to 0. Normally many more ferrite cores are wired together, as in Fig. 10-7, to form a memory plane.

Next we shall look at the read operation in a coincident-current memory. Figure 10-8 shows four ferrite cores wired with the write wires (X_1, X_2, Y_1, Y_2). A third wire, called the *sense wire*, is added to the matrix. The sense wire is shown in Fig. 10-8. To read the contents of core 4, a $-\frac{1}{2}I$ pulse is applied to the X_2 and Y_2 lines. Core 4 is reset to 0 by the addition of these two currents. As it changes magnetic states (from ccw to cw), core 4 *induces* a pulse in the sense line, which tells us we have a 1 in core 4. Core 1 receives no current. Cores 2 and 3 receive only half the current needed to reset them to 0. Cores 2 and 3 do not change magnetic state. Only the core that was addressed (core 4) can change state and trigger the sense wire.

A group of ferrite cores wired somewhat like

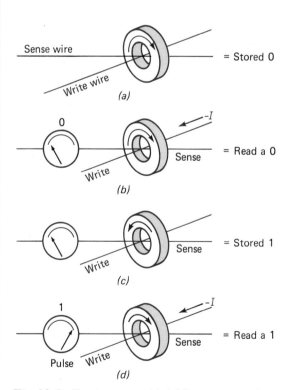

Fig. 10-6   Ferrite core. (*a*) **Adding a sense wire.** (*b*) **Reading a 0.** (*c*) **Storing a 1.** (*d*) **Reading a 1.**

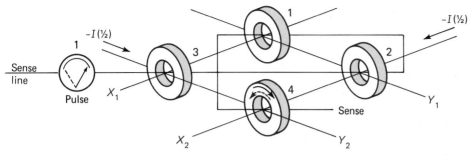

Fig. 10-8 Using a sense line to read a coincident-current core memory.

the one in Fig. 10-8 forms a *plane*. One plane contains many cores. Figure 10-9 shows how 64 cores are arranged in a plane. In this arrangement four planes are stacked on top of one another. This would form a 64 × 4-bit memory. This 256-bit magnetic core memory would have 64 words, each word would be 4 bits long. Figure 10-9 shows how cores X_6Y_5 on all four planes are located. This is Word X_6Y_5. The dots on each plane represent the cores being addressed by the X and Y select lines. The lines shown are the write lines we mentioned earlier. Notice that these write (X and Y select) lines are wired in series from plane to plane.

Sense lines from each board (or plane) would be brought out for a parallel readout of the four planes. The sense outputs are represented on the right. The sense wires are threaded through the ferrite cores parallel to the X select lines.

A fourth wire, called the *inhibit wire*, is threaded through each ferrite core. The inhibit wires run through the cores parallel to the Y select lines. The inhibit leads for each plane are represented on the left side of Fig. 10-9.

Suppose we are writing a 1 into the Word X_6Y_5 as shown in Fig. 10-9. This means that binary 1111 is written into the four cores. What if we wanted to write 1011 ($D = 1$, $C = 0, B = 1, A = 1$) instead? We simply activate the C plane inhibit line; it cancels the command from the select (write) lines. The inhibit lines can be thought of as a means to write a 0 in a word.

The 256-bit magnetic core memory shown in Fig. 10-9 is a nonvolatile memory. This memory still needs decoding for selecting the correct address. This memory destroys its contents when reading. Schemes for writing

contents back in the cores after the reading process is complete are available. You can imagine that a core memory of this type is difficult and expensive to manufacture.

Magnetic core memories are commonly used as internal central storage in large computers. Magnetic core memories may be replaced in time by less expensive units such as semiconductor or other types of memories. Until this happens, however, the ferrite core will remain the memory cell of the digital computer.

10-8 COMPUTER BULK STORAGE DEVICES

Semiconductor and magnetic core memories are used for the internal storage in computers. It is not possible to store all data inside the computer itself. For instance, it is not necessary to store last month's payroll information inside the computer after the checks are printed and cashed. Thus most data is stored outside the computer. Several methods are used to store information for immediate and future use by a computer. External storage devices can be classified as *mechanical* or *magnetic*.

Mechanical bulk storage devices are the familiar punched paper card and punched or perforated paper tape. The punched card is an extremely common method of storing information. Most information is put on punched cards at one time or another. Perforated paper tape is a narrow strip of paper, with holes punched across the tape at places selected according to a code. The paper tape can be stored on reels.

Common magnetic bulk storage devices are the magnetic tape, the magnetic disc, and the magnetic drum. Each device operates very

105

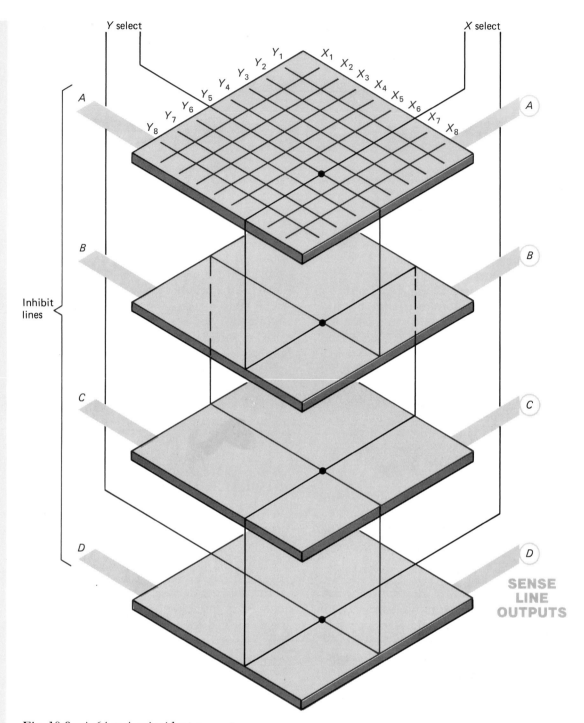

Fig. 10-9   A 64 × 4 coincident-current core memory.

much like a common tape recorder. Information is recorded (stored) on the magnetic material. Information can also be read from the magnetic material.

Most equipment you see in a computer room are peripheral devices. Peripheral de-

vices are not part of the computer; they only feed information into or take information out of the computer. Many of these peripheral devices are used just to handle punched cards, perforated tape, magnetic tape, magnetic drums, and magnetic discs.

106

Summary

1. Common semiconductor memories are RAMs, ROMs, PROMs, and read-mostly memories.

2. A RAM is considered a read/write random-access memory device.

3. A ROM is considered a permanent storage unit that has the read-only characteristic.

4. A PROM operates just like a ROM. PROMs are one-time write devices. A read-mostly memory is a reprogrammable ROM.

5. The write process stores information in the memory. The read or sense process detects the contents of the memory.

6. An integrated circuit RAM can be reprogrammed easily but is a volatile memory.

7. A 64 × 4 memory will hold 64 words each 4 bits long. It will hold a total of 256 bits of data.

8. The magnetic core memory is based upon the magnetic characteristics of the ferrite core for storing 0s and 1s.

9. Ferrite cores are organized into planes, and the planes are stacked. Read/write, sense, and inhibit lines are threaded through tiny cores in a magnetic core coincident-current memory.

10. Computer external storage methods are magnetic tapes, discs, and drums; punched cards; and perforated tape.

Questions

1. Press the store key on a calculator. This activates the ___?___ (read, write) process in the memory section.

2. Press the recall key on a calculator. This activates the ___?___ (read, write) process in the memory section.

3. The following abbreviations stand for what?
   a. RAM
   b. ROM
   c. PROM

4. A ___?___ (RAM, ROM) has volatile memory.

5. A ___?___ (RAM, ROM) has both the read and write capability.

6. A ___?___ (RAM, ROM) is a permanent memory.

7. A ___?___ (RAM, PROM) is a nonvolatile memory.

8. A ___?___ (RAM, ROM) would have a read/write input control.

9. A ___?___ (RAM, ROM) would have data inputs.

10. A RAM such as the 7489 IC is sometimes also called by what two other names?

11. A 32 × 8 memory could hold ___?___ words. Each word is ___?___ bits long.

12. Draw a diagram of how a 32 × 8 memory would look in table form. The diagram will be similar to Fig. 10-1.

13. Draw a logic diagram of a ROM organized as a 32 × 8 storage unit. Label address inputs as E, D, C, B, and A. Label outputs as DO_1, DO_2, DO_3, DO_4, DO_5, DO_6, DO_7, and DO_8.

14. How would you program a ROM such as the one in questions 12 and 13?

15. List at least three advantages of semiconductor memories.

16. A ___?___ (RAM, ROM) can be erased easily.

17. List at least three uses of ROMs.

18. What two methods are used for erasing a read-mostly memory?

19. The ferrite core is the memory cell in a ____?____-type memory.

20. The write wire through a ferrite core is used for ____?____ and reading.

21. The sense wire through a ferrite core is used during the ____?____ (read, write) process.

22. In a ferrite core, if magnetic flux travels in a clockwise direction for a logical 0, then it will travel ____?____ for a logical 1.

23. A ____?____ (magnetic core, RAM) unit has a nonvolatile memory.

24. A coincident-current-type core memory uses four wires threaded through the cores. Name the wires.

25. List at least four common types of computer bulk (external) storage.

Digital Systems

■ Most digital devices we use every day are *digital systems,* such as hand-held calculators, digital wristwatches, or even digital computers. Calculators, digital clocks, and computers are an assembly of *subsystems.* Typical subsystems might be adders/subtractors, counters, shift registers, RAMs, ROMs, encoders, decoders, data selectors, clocks, and display decoder/drivers. You have already used most of these subsystems. This chapter discusses the various digital systems and how they transmit data. A digital system is formed by the proper assembly of digital subsystems.

11-1 ELEMENTS OF A SYSTEM

Most mechanical, chemical, fluid, or electrical systems have certain features in common. Systems have an *input* and an *output* for their product, power, or information. Systems also act on the product, power, or information; this is called *processing.* The entire system is organized and its operation directed by a *control* function. The *transmission* function transmits products, power, or information. More complicated systems also contain a *storage* function. Figure 11-1 illustrates the overall organization of a system. Look carefully and you can see that this diagram is general enough to apply to nearly any system, whether it is transportation, fluid, school, or electronic. The transmission from device to device is shown by the colored lines

and arrows. Notice that the data or whatever is being transferred always moves in one direction. It is common to use double arrows on the control lines to show that the control unit is directing the operation of the system as well as receiving feedback from the system.

The general system shown in Fig. 11-1 will help explain several digital systems in this chapter. In a digital system we shall be dealing only with transmitting data (numbers).

11-2 A DIGITAL SYSTEM ON AN IC

We learned that all digital systems could be wired from individual AND and OR gates and inverters. And we learned that manufacturers produced subsystems on a single IC (counters, registers, and so on). We shall find that manufacturers have gone even a step fur-

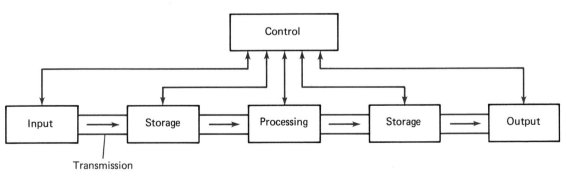

Fig. 11-1 The elements of a system.

ther: some ICs contain nearly an entire digital system.

Texas Instruments defines the least complex digital ICs as *small-scale integrations* (SSI). An SSI will contain circuit complexity up to about 10 equivalent gates or circuitry of similar complexity. Small-scale ICs include the gate and flip-flop ICs you used.

A *medium-scale integration* (MSI) has the complexity of from 12 to 100 equivalent gates. ICs that are classified as MSIs belong to the subsystem group. Typical examples are adders, shift registers, code converters, counters, data selectors/multiplexers, RAMs, and ROMs. Most of the ICs you have studied and used to this time have been SSIs or MSIs.

A *large-scale integration* (LSI) has the complexity of more than 100 equivalent gates. A major subsystem or an entire digital system is fabricated in a single IC. Examples are digital clock ICs, calculator ICs, and microprocessor ICs. Large-scale ICs can be considered a digital system on a *chip*. The term chip refers to the single silicon wafer that contains all the electronic circuitry inside an IC.

11-3 THE CALCULATOR

The pocket calculator in nearly everyone's pocket or desk is a very complicated digital system. Knowing this, it is disappointing to take apart a modern miniature calculator. You will find a battery, the tiny readout displays, a few wires from the keyboard, and a circuit board with an IC attached. That single IC is most of the digital system we call a calculator and contains an LSI chip that performs the task of hundreds and thousands of logic gates. The single IC performs the storage, processing, and control functions of the calculating system. The keyboard is the input, and the displays are the output of the calculator system.

What happens inside the calculator chip when you press a number or add two numbers? The diagram in Fig. 11-2 will help us figure out how a calculator works. Figure 11-2 shows three components: the keyboard, the seven-segment displays, and the power supply. These parts are the only functional ones *not* contained in the single LSI IC in most small calculators. The keyboard is obviously the input device. The keyboard contains simple, normally-open switches. The decimal display is the output. The readout unit in Fig. 11-2 contains only six seven-segment displays. The power supply is a battery in most inexpensive hand-held calculators.

The calculator chip (IC) is divided into several functional subsystems, as shown in Fig. 11-2. The organization shown is only one of several ways to get a calculator to operate. The heart of the system is the adder/subtractor subsystem, which operates very much like the 4-bit adders you studied. The clock subsystem pulses all parts of the system at a constant frequency. The clock frequency is fairly high; ranging from 25 to 500 kHz. When the calculator is turned on, the clock runs constantly, and the circuits "idle" until a command comes from the keyboard.

Suppose we add $2 + 3$ with this calculator. As we press the 2 on the keyboard, the encoder translates the 2 to a BCD 0010. The 0010 is directed to the display register by the control circuitry and is stored in the display register. This information is also applied to the seven-segment decoder, and lines a, b, d, e, and g are activated. The first (1s display) seven-segment display shows a 2 when the scan line pulses that unit briefly. This scanning continues at a high frequency, and the display appears to be lit continuously, even though it is being turned on and off many times per second. Next, we press $+$ on the keyboard. This operation is transferred to and stored in code form in an extra register (X register). Now we press the 3 on the keyboard. The encoder translates the 3 to a BCD 0011. The 0011 is transferred to the display register by the controller and is placed in the display register, which also places a 3 on the display. Meanwhile, the controller has moved the 0010 (2) to the operand register. Now we press the $=$ key. The controller checks the extra register to see what to do. The X register says to add the BCD numbers in the operand and display registers. The controller applies the contents of the display and operand registers to the adder inputs. The results of the addition collect in the accumulator register. The result of the addition is a BCD 0101. The controller routes the answer to the display register, shown on the readout as a 5.

For longer and more complex numbers containing decimal points, the controller will follow directions in the instruction register. For complicated problems the unit may cycle through hundreds of steps as programmed

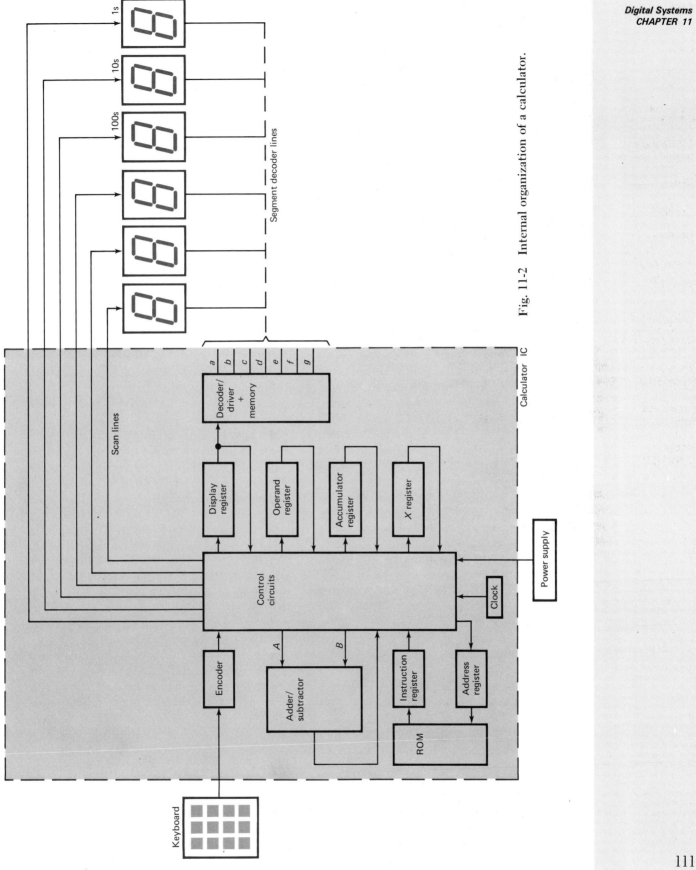

Fig. 11-2 Internal organization of a calculator.

into the ROM. But amazingly, even hundreds of operations take less than $\frac{1}{10}$ s.

The shift registers in Fig. 11-2 are rather large units compared to the ones you used in the lab. The ROM also has a large capacity (many thousands of bits). The calculator in Fig. 11-2 is but one example of how a calculator can operate. Each commercial unit operates in its own unique way. This discussion serves only to point out that many subsystems you have already used are found in a complex digital system like the calculator.

Only the original IC designers needed to know the organization of the subsystems in Fig. 11-2. This organization is sometimes referred to as the *architecture* of the calculator. Notice that all the elements of a system are present in this electronic calculator.

11-4 THE COMPUTER

The most complex digital systems are *computers*. Most computers can be divided into the five functional sections shown in Fig. 11-3. The input device may be a keyboard, card reader, magnetic tape unit, or telephone line. This equipment lets us pass information from *person to machine*. The input device must *encode* human language into the binary language of the computer.

The memory section is the storage area for data and program. This storage can be supplemented by storage outside the processing unit. Much of the memory in the *central processing unit* (CPU) traditionally has been magnetic core memory, but now semiconductor memories are also being used in the CPU.

The arithmetic unit is what most people think of as being inside a computer. The arithmetic unit adds, subtracts, compares, and does other logic functions. Notice that a two-way path exists between the memory and arithmetic sections. In other words, data can be sent to the arithmetic section for action and the results sent back to storage in the memory. The arithmetic unit is sometimes referred to as the arithmetic logic unit (ALU).

The control section is the nervous system of the computer. It directs all other sections to operate in the proper order and tells the input when and where to place information in the memory. It directs the memory to route information to the arithmetic section and tells the arithmetic section to add. It routes the answer back to the memory and routes the answer to the output device. It tells the output device when to operate. This is only a sampling of what the control section can do.

The output section is the link between the *machine and a person*. It can communicate to humans through a printer, which is something like a typewriter without the keys. It can put out information on a televisionlike cathode-ray-tube display. Output information can also be placed on bulk storage devices such as punched cards or magnetic tape. The output section must *decode* the language of the computer into human language.

The entire center section in Fig. 11-3 is often called the CPU. The arithmetic, memory, and most of the control section are frequently housed in a single cabinet. Devices located outside the CPU are often called *peripheral devices*.

The block diagram of the computer in Fig.

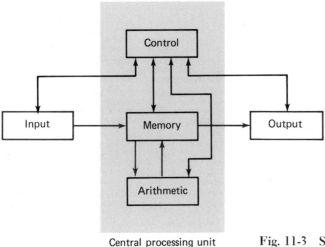

Central processing unit

Fig. 11-3 Sections of a digital computer.

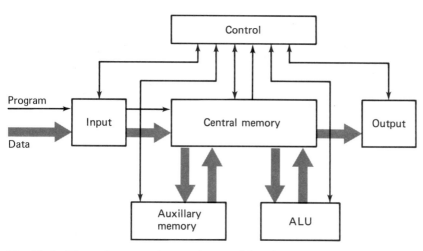

Fig. 11-4   Flow of program instructions and data in a computer.

11-3 could well be the diagram for a calculator. Up to this point the basic systems operate the same. The basic difference between the calculator and computer is *size* and the use of a *stored program* in the computer. Figure 11-4 shows that two types of information are put into the computer. One is the program (instructions) telling the control unit how to proceed in solving the problem. This program, which has to be carefully written by a programmer, is stored in the central memory while the problem is being solved. The second type of information fed to the computer is *data*, to be acted on by the computer. Data are the facts and figures needed to solve the problem. Notice that the program information is placed in storage in the memory and used only by the control unit. The data information, however, is directed to various positions within the computer and is processed by the ALU. The data need never go to the control unit. The auxiliary memory is extra memory that may be needed to store the vast amount of data in some complex problems. It may not be in the CPU. Data may be stored in peripheral devices.

In summary, the computer is organized into five basic functional sections: the input, memory, control, arithmetic/logic unit, and output. Information fed into the CPU is either program instructions or data to be acted upon. The computer's stored program and size make it different from the calculator.

Computers, the most complex of digital systems, were not covered in depth in this unit. There are entire books about the organization and architecture of computers. Remember, however, that all the circuits in the computer are constructed from logic gates, flip-flops, and subsystems such as the ones you have studied.

11-5 DATA TRANSMISSION

Most data in digital systems is transmitted directly through wires and printed circuit boards. Many times bits of data must be transmitted from one place to another. Sometimes the data must be transmitted over telephone lines or cables to points far away. If all the data were sent at one time over *parallel* wires, the cost and size of these cables would have to be too expensive and large. Instead, the data are sent over a single wire in *serial* form and reassembled into parallel data at the receiving end. The devices used for sending and receiving serial data are called *multiplexers* (MUX) and *demultiplexers* (DEMUX).

The basic idea of a *multiplexer* and *demultiplexer* is shown in Fig. 11-5. Parallel data from one digital device are changed into *serial* data by the multiplexer. The serial data are transmitted by a single wire. The serial data are reassembled into parallel data at the output by the demultiplexer. Notice the control lines that must also connect the multiplexer and demultiplexer. These control lines keep the multiplexer and demultiplexer synchronized. Notice that the 16 input lines are cut down to only a few transmission lines.

The system in Fig. 11-5 works in the following manner. The multiplexer first will connect input 0 to the serial data transmission line. The bit will be transmitted to the de-

113

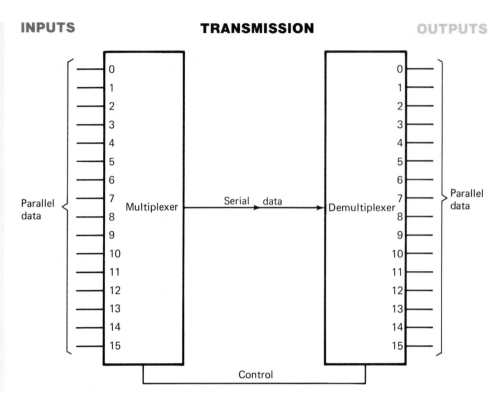

Fig. 11-5   Serial data transmission using a multiplexer and demultiplexer.

multiplexer, which will place this bit of data at output 0. The multiplexer and demultiplexer proceed to transfer the data at inputs 1 to 3 to outputs 1 to 3 and so on. The bits are transmitted 1 bit at a time.

A *multiplexer* works much like a single-pole, many-position rotary switch, as shown in Fig. 11-6. Rotary switch SW 1 shows the action of a multiplexer. The demultiplexer operates like rotary switch SW 2 in Fig. 11-6. The mechanical control in this diagram makes sure input 5 on SW 1 is delivered to output 5 on SW 2. Notice that the mechani-

cal switches in Fig. 11-6 will permit data to travel in either direction. Being made from logic gates, multiplexers and demultiplexers will permit data to travel only from input to output, as in Fig. 11-5.

You used a multiplexer before, in Chap. 4. The other name for a multiplexer is a *data selector*. The *demultiplexer* is sometimes called a *distributor* or *decoder*. The term distributor describes the action of SW 2 in Fig. 11-6 as it "distributes" the serial data first to output 1, then to output 2, then to output 3, and so forth.

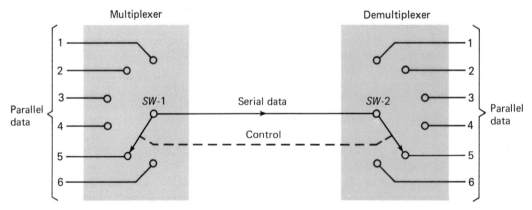

Fig. 11-6   Rotary switches act like multiplexers and demultiplexers.

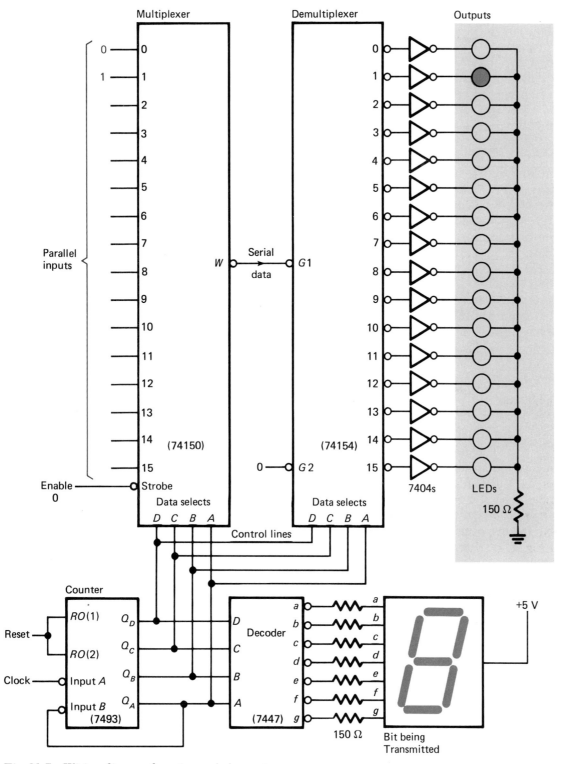

Fig. 11-7 **Wiring diagram for a transmission system.**

Figure 11-7 is a detailed wiring diagram of a transmission system using the multiplexer-demultiplexer arrangement. A word (16 bits long) is entered at the inputs (0 to 15) of the 74150 multiplexer IC. The 7493 counter starts at binary 0000. This is shown as 0 on the seven-segment display. With the data select inputs (D, C, B, A) of the 74150 multiplexer at 0000, it will take the data from input 0, which is shown as a logical 0. The logical 0 is transfered to the 74154 demultiplexer IC, where it is routed to output 0. Normally the output of the 74154 IC would be inverted as shown by the invert bubbles. A 7404 inverter complements the output 0 back to the original logical 0.

The counter increases to binary 0001. This is shown as a 1 on the decimal readout. This binary 0001 is applied to the data select inputs of both ICs (74150 and 74154). The logical 1 at the input of the 74150 multiplexer is transferred to the transmission line. The 74154 demultiplexer routes the data to output 1. The 7404 inverter complements the output, and the logical 1 appears as a lighted LED, as shown in the diagram. The counter continues to scan each input of the 74150 IC and transfer the contents to the output of the 74154 demultiplexer. Notice that the counter must count from binary 0000 to 1111 (16 counts to transfer just one parallel word from the input to output of this system. The seven-segment LED readout provides a convenient way of keeping track of which input is being transmitted. If the clock were pulsed very fast, the parallel data could be transmitted quite quickly as serial data to the output.

Notice from Fig. 11-7 that we have saved many pieces of wire by sending the data in *serial* form. This takes somewhat more time, but the rate we send data over the transmission line can be very high.

11-6 DETECTING ERRORS IN DATA TRANSMISSIONS

Digital equipment such as a computer is valuable to people because it is fast and *accurate*. To help make digital devices accurate, special *error-detection* methods are used. You can

Table 11-1 Truth table for parity bit generator.

INPUTS			OUTPUT
Parallel data			Parity bit
C	B	A	P
0	0	0	0
0	0	1	1
0	1	0	1
0	1	1	0
1	0	0	1
1	0	1	0
1	1	0	0
1	1	1	1

well imagine an error creeping into a system when data are transferred from place to place.

To detect errors we must keep a constant check on the data being transmitted. To check accuracy we generate and transmit an extra *parity bit*. Figure 11-8 shows such a system. In this system three parallel bits (A, B, C) are being transmitted over a long distance. Near the input they are fed into a *parity-bit generator* circuit. This circuit generates what is called a parity bit. The parity bit is transmitted with the data, and near the output the results are checked. If an error occurs during transmission, the *error-detector* circuit sounds an alarm. If all the parallel data are the same at the output as they were at the input, no alarm sounds.

Table 11-1 will help you understand how the error-detection system works. This table is really a truth table for the parity-bit generator in Fig. 11-8. Notice that the inputs are labeled A, B, and C for the three data transmis-

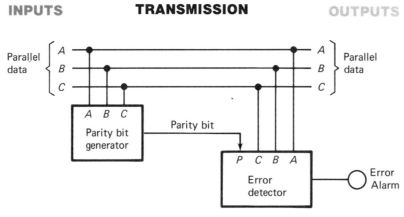

Fig. 11-8 Error-detection system using a parity bit.

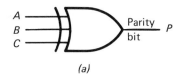

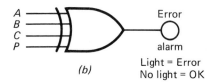

Fig. 11-9 (a) Parity-bit generator circuit. (b) Error-detector circuit.

sion lines. The output was determined by looking across a horizontal row. We want an *even number of 1s* in each row (zero 1s, or two 1s, or four 1s). Notice that row 1 has no 1s. Row 2 has a single 1 plus the parity bit 1. Row 2 now has two 1s. As you look down Table 11-1 you will notice that each horizontal row contains an even number of 1s. Next, the truth table is converted into a logic circuit. The logic circuit for the parity-bit generator is drawn in Fig. 11-9(*a*). You can see that a 3-input XOR gate will do the job for generating a parity bit. The 3-input XOR gate in Fig. 11-9, then, would be the logic circuit you

would substitute for the parity-bit generator block in Fig. 11-8.

Look at the entire truth table in Table 11-1. We see that under normal circumstances each horizontal row contains an *even number* of 1s. Were an error to occur, we would then have an *odd number* of 1s appear. A circuit that gives a logical 1 output any time an odd number of 1s appear is shown in Fig. 11-9(*b*). A 4-input XOR gate would detect an odd number of 1s at the inputs and turn on the alarm light. Figure 11-9(*b*) diagrams the logic circuit that would substitute for the error-detector block in Fig. 11-8.

The use of the parity bit only warns you of an error; the system we used did *not* correct the error. Some codes that are *error-correcting*, such as the *Hamming code*, have been developed. The Hamming code uses several extra parity bits when transmitting data.

11-7 ADDER/SUBTRACTOR SYSTEM

In Chap. 9 you worked with adders and subtractors and studied an adder/subtractor system. Figure 11-10 is a block diagram of that system (the complete wiring diagram is shown in Fig. 9-18).

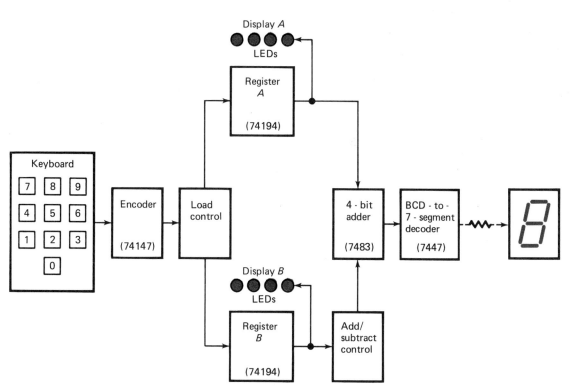

Fig. 11-10 Block diagram of an adder/subtractor system.

The system diagramed in Fig. 11-10 is made up of familiar subsystems. The keyboard is the input device to this system. An encoder (74147 IC) converts the keyboard input to binary. The load control routes data to either storage unit (register A or register B). Registers A and B hold data at the inputs of the 4-bit adder while the calculation is performed. For subtraction the add/subtract control unit performs the 1s complement and end-around carry procedure. The sum from the 7483 IC adder is applied to the decoder (7447). The decoder translates from binary to the seven-segment code. The seven-segment display reads out the sum or difference in decimal. Displays A and B show in binary the contents at the outputs of the two 74194 shift registers.

11-8 THE DIGITAL CLOCK

We introduced a digital electronic clock in Chap. 7 and noted that various *counters* are the heart of a digital clock system. Figure 11-11(*a*) is a simple block diagram of a digital clock system. Most clocks use the power line frequency of 60 Hz as their input. This frequency is divided into seconds, minutes, and hours by the *frequency divider* section of the clock. The one-per-second, one-per-minute, and one-per-hour pulses are then counted and stored in the *count accumulator* section of the clock. The stored contents of the count accumulators (seconds, minutes, hours) are then *decoded*, and the correct time is shown on the output *time displays*. The digital clock has the typical elements of a system. The input is the 60-Hz alternating current. The processing takes place in the frequency divider, count accumulator, and decoder sections. The storage takes place in the count accumulator. The control section is illustrated by the *time-set* control, as shown in Fig. 11-11(*a*). The output section is the digital time display.

It was mentioned that all systems are made up of logic gates, flip-flops, and subsystems. The diagram in Fig. 11-11(*b*) shows how subsystems are organized to display time in hours, minutes, and seconds. This is a more detailed diagram of a digital clock. The input is still a 60-Hz signal. The 60 Hz is from the low-voltage secondary of a transformer. The 60 Hz is divided by 60 by the first frequency divider. The output of the first divide-by-60

circuit is one pulse per second. The one pulse per second is put into an *up counter* that counts upward from 00 through 59 and then resets to 00. The seconds' counters are then decoded and displayed on the two seven-segment LED displays at the upper right, Fig. 11-11(*b*).

Consider the middle frequency-divider circuit in Fig. 11-11(*b*). The input to this divide-by-60 circuit is one pulse per second; the output is one pulse per minute. The one-pulse-per-minute output is transferred into the 0 to 59 minutes' counter. This up counter keeps track of the number of minutes from 00 through 59 and then resets to 00. The output of the minutes' count accumulator is decoded and displayed on the two seven-segment LEDs at the top center, Fig. 11-11(*b*).

Now for the divide-by-60 circuit on the right in Fig. 11-11(*b*). The input to this frequency divider is one pulse per minute. The output of this circuit is one pulse per hour. The one pulse per hour is transferred to the hours' counter on the left. This hours' count accumulator keeps track of the number of hours from 0 to 23. The output of the hours' count accumulator is decoded and transferred to the two seven-segment LED displays at the upper left, Fig. 11-11(*b*). You probably have noticed already that this is a 24-h digital clock. It easily could be converted to a 12-h clock by changing the 0 to 23 count accumulator to a 0 to 11 counter.

For setting the time a time-set control has been added to the digital clock in Fig. 11-11(*b*). When the switch is closed (logic gate may be used), the display will count forward at a fast rate. This enables you to set the time quickly. The switch bypasses the first divide-by-60 frequency divider so the clock moves forward at 60 times its normal rate. An even faster *fast-forward* set could be used by bypassing both the first and second divide-by-60 circuits. The latter technique is common in digital clocks.

What is inside the divide-by-60 frequency dividers in Fig. 11-11(*b*)? In Chap. 7 we spoke of a counter being used to divide frequency. Figure 7-9 illustrates a modulo-6 and a decade counter connected. These counters form a divide-by-60 circuit that will work in our digital clock.

The seconds' and minutes' count accumulators in Fig. 11-11(*b*) are also counters. The 0 to 59 counter is a decade counter cascaded

INPUT   OUTPUTS

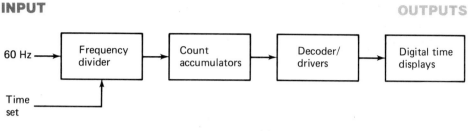

(a)

OUTPUT

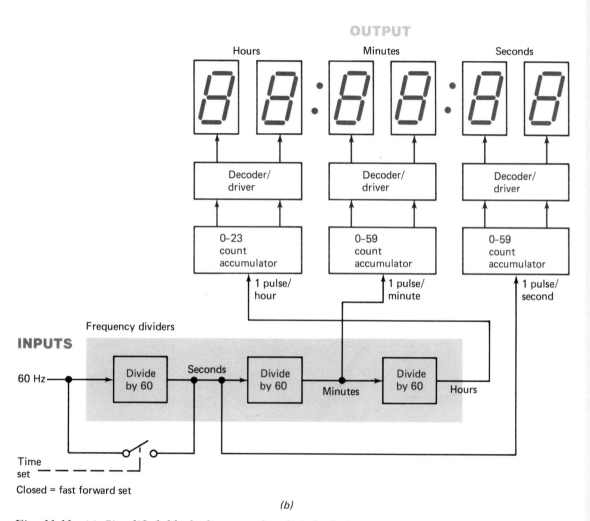

(b)

Fig. 11-11 *(a)* Simplified block diagram of a digital clock.
(b) More detailed block diagram of a digital clock.

with a 0 to 5 counter. The decade counter is coupled with the 1s place of the displays. The modulo-6 counter is coupled with the 10s place of the displays. In a like manner, the hours' count accumulator is a decade counter cascaded with a 0 to 2 counter. The decade counter is coupled to the 1s place in the hours' display. The modulo-3 counter is coupled with the 10s place of the hours' display.

In many practical digital clocks the output may be in hours and minutes only. Most digital clocks are based upon one of many inexpensive ICs. The *clock chips* (LSI) will have all the frequency dividers, count accumulators, and decoders built into the single IC. For only a few dollars, clock chips have other features, such as 12- or 24-h outputs, calendar features, alarm controls, and radio controls.

Wave-shaping
circuit

One-shot
multivibrator

Pulse width

Frequency
counter

An added feature you will use when you construct a digital timepiece is shown in Fig. 11-12(a). A *wave-shaping circuit* has been added to the block diagram of our digital clock. The IC counters that make up the frequency-divider circuit do not work well with a sine-wave input. The sine wave [shown at the left in Fig. 11-12(a)] has a slow *rise time* that will not trigger the counter properly. The sine-wave input must be converted into a square wave. The wave-shaping circuit changes the sine wave to a square wave. The square wave will now properly trigger the frequency-divider circuit.

Commercial clock chips (LSI) have this wave-shaping circuit built into the IC. In the lab you will use a *one-shot multivibrator* (monostable multivibrator) to square up the input sine wave. A one-shot multivibrator puts out a single square-wave pulse for a short time when the input is triggered. One-shot multivibrators can be wired using discrete components. In the lab you will probably use an IC that contains a one-shot multivibrator. Figure 11-12(b) shows a 74121 one-shot multivibrator IC. The 74121 is wired to form a square-wave output with a sine-wave input. The 33-kΩ resistor and the 0.01 μF capacitor determine the *pulse width* of the output pulse. The pulse width is the time the pulse stays high. The pulse width in this circuit is

about 100 μs. In the wave-shaping circuit in Fig. 11-12(b), when the ac input reaches about 1.7 V positive at input B (compared to GND), the 74121 triggers, giving a single pulse. The pulse is about 100 μs long. A single short square-wave pulse appears at the output of the 74121 each time the sine wave goes positive to about 1.7 V. The 74121 one-shot multivibrator in Fig. 11-12(b) also has a control that can start and stop the operation of the multivibrator. A logical 0 on the control enables the multivibrator to operate normally; a logical 1 on the control inhibits the operation of the unit.

You will want to get some practical knowledge of how counters are used in dividing frequency. Remember that the counter subsystem will be used for two jobs in the digital stopwatch: first to divide frequency, and second to count upwards and keep track of the number of pulses at its input.

11-9 THE FREQUENCY COUNTER

An instrument used by technicians and engineers is the *frequency counter*. A digital frequency counter shows in decimal numbers the frequency in a circuit. Counters can measure from low frequencies of a few cycles per second (hertz, Hz) up to very high fre-

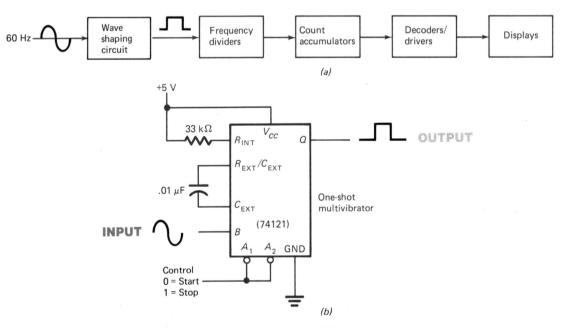

Fig. 11-12 (a) Adding a wave-shaping circuit to the input of the digital clock system. (b) A 74121 one-shot multivibrator being used as a wave-shaping circuit.

quencies of thousands of megahertz (MHz). Like a digital clock, the frequency counter uses decade counters as its heart.

As a review, the block diagram for a digital clock is shown in Fig. 11-13(a). The known frequency is divided properly by the counters in the clock. The counter outputs are decoded and displayed in the time display. Figure 11-13(b) shows a block diagram of a frequency counter. Notice that the frequency counter circuit is fed an *unknown* frequency instead of the known frequency in a digital clock. The counter circuit in the frequency counter in Fig. 11-13(b) also contains a *start/stop control*.

The frequency counter has been redrawn in Fig. 11-14(a). Notice that an AND gate has been added to the circuit. The AND gate will control the input to the decade counters. When the start/stop control is at a logical 1, the unknown frequency pulses go through the AND gate and on to the decade counters. The counters count upward until the start/stop control returns to a logical 0. The 0 turns off the control gate and stops the pulses from getting to the counters.

Figure 11-14(b) is a more exact timing diagram of what happens in the frequency

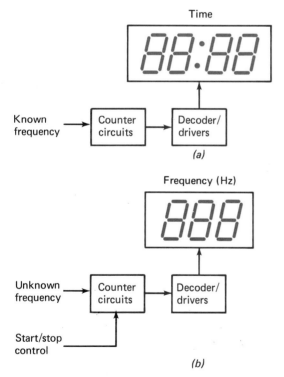

Time

Known frequency → **Counter circuits** → **Decoder/ drivers**

(a)

Frequency (Hz)

Unknown frequency → **Counter circuits** → **Decoder/ drivers**

Start/stop control

(b)

Fig. 11-13 (a) Simplified block diagram of a digital clock. (b) Simplified block diagram of a digital frequency counter.

counter. Line A shows the start/stop control at logical 0 on the left and then going to 1 for *exactly one second*. The start/stop control then returns to logical 0. Line B diagrams a continuous string of pulses from the unknown frequency input. The unknown frequency and the start/stop control are ANDed together as we saw in Fig. 11-14(a). Line C in Fig. 11-14(b) shows only the pulses that are allowed through the AND gate. These pulses trigger the up counters. Line D shows the count that would be observed on the displays. Notice that the displays start cleared to 00. The displays then count upward to 11 during the one second. The unknown frequency in line B in Fig. 11-14(b) is shown as 11 Hz (11 pulses per second).

A somewhat higher frequency is put into the frequency counter in Fig. 11-14(c). Again line A shows the start/stop control beginning at 0. It is then switched to a logical 1 for *exactly one second*. It is then returned to logical 0. Line B in Fig. 11-14(c) shows a string of higher frequency pulses. This is the unknown frequency being measured by this digital frequency counter. Line C shows the pulses that trigger the decade counters during the one second count-up period. The decade counters sequence upward to 19, as shown in line D. The unknown frequency in Fig. 11-14(c) is then 19 Hz (19 pulses in 1 s).

If the unknown frequency were 870 Hz, the counter would count from 000 to 870 during the one second count period. The 870 would be displayed for a time, and then the counters would be reset to 000 and the frequency would be counted again. This *reset-count-display sequence* is repeated over and over.

Notice that the start/stop control pulse (count pulse) must be *very accurate*. Figure 11-15 shows how a count pulse can be generated by using an accurate known frequency such as the 60 Hz from the power line. The 60-Hz sine wave is converted into a square wave by the wave-shaping circuit. The 60-Hz square wave is run through a counter that divides the frequency by 60. The output is a pulse *exactly one second* in length. This *count pulse* "turns on" the control circuit when it goes high and permits the unknown frequency to trigger the counters. The unknown frequency is applied to the counters for exactly one second.

Remember that the frequency counter goes through the reset-count-display sequence.

Reset-counter-
display
sequence

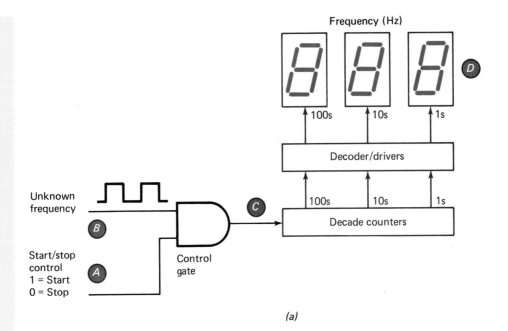

(a)

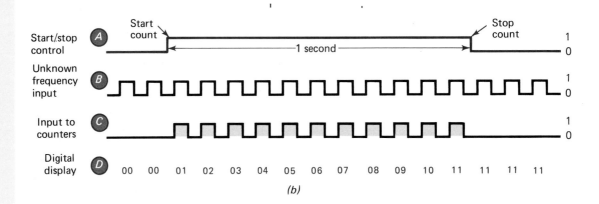

(b)

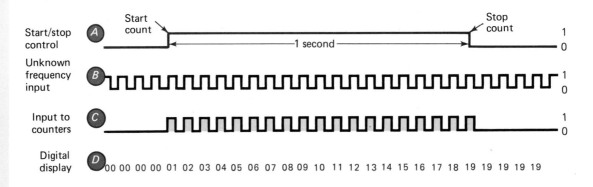

(c)

Fig. 11-14 (*a*) Block diagram of a digital frequency counter showing the start/stop control. (*b*) Waveform diagram for an unknown frequency of 11 Hz. (*c*) Waveform diagram for an unknown frequency of 19 Hz.

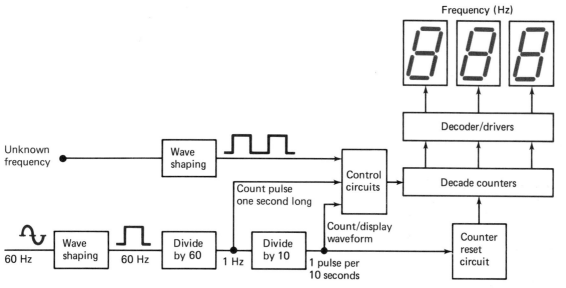

Fig. 11-15 More detailed block diagram of a digital frequency counter.

So far we have shown only the count part of this sequence. The *counter-reset* circuit is a group of gates that resets or clears the decade counters to 000 at the correct time—just before the count starts. Next, the one second count pulse permits the counters to count upward. The count pulse ends, and the unknown frequency is *displayed* on the seven-segment displays. In this circuit, the frequency will read in hertz. It is convenient to leave this display on the LEDs for a time. To do this, the divide-by-10 counter sends a pulse to the control circuit, which "turns off" the count sequence for nine seconds. Events then happen like this. *Reset* the counters to 000. *Count* upward for one second. *Display* the unknown frequency for nine seconds with no counts. Repeat the reset-count-display procedure every ten seconds.

The frequency counter in Fig. 11-15 will measure frequencies from 1 to 999 Hz. Notice the extensive use of counters in the divide-by-60, divide-by-10, and three decade counter circuits—hence the name frequency counter. The digital frequency counter actually counts the pulses in a given amount of time.

One limitation of the counter diagramed in Fig. 11-15 is its top frequency; the top frequency that could be measured was 999 Hz. There are two ways to increase the top frequency of our counter. The first method is to add one or more counter-decoder-display units. We could extend the range of the fre-

quency counter in Fig. 11-15 to a top limit of 9999 Hz by adding a counter-decoder-display unit.

The second method of increasing the frequency range is to count by 10s instead of 1s. This idea is illustrated in Fig. 11-16. A divide-by-6 counter has replaced the divide-by-60 unit in our former circuit. This makes the *count pulse* only *one-tenth* of a second long. The count pulse will permit only one-tenth as many pulses through the control as with the one second pulse. This is the same as counting by 10s. Only three LED displays are used. The 1s display in Fig. 11-16 is only to show that a 0 must be added to the right of the three LED displays. The range of this frequency counter would be from 10 to 9990 Hz.

In the circuit in Fig. 11-16, the decade counters count upward for $\frac{1}{10}$ s. The display is held on the LEDs for $\frac{9}{10}$ s. The counters are then reset to 000. The count-display-reset procedure is then repeated. The circuit in Fig. 11-16 has one other new feature: during the count time the displays are blanked out in this circuit. They are then turned on again when the unknown frequency is on the display. The sequence for this frequency counter is then reset, count (with displays blank), and, finally, the longer display period. This sequence repeats itself every one second while the instrument is being used.

The frequency counter diagramed in Fig. 11-16 is similar to one you can assemble in

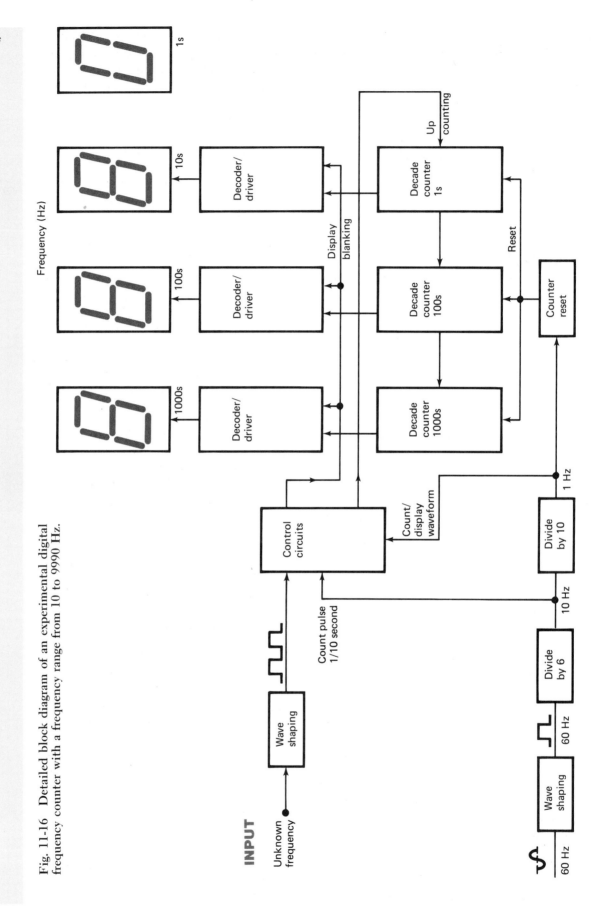

Fig. 11-16 Detailed block diagram of an experimental digital frequency counter with a frequency range from 10 to 9990 Hz.

the lab from gates, flip-flops, and subsystems. It is highly suggested that you set up this complicated digital system because only practical experience will teach you the details of the frequency counter system.

11-10 AN EXPERIMENTAL FREQUENCY COUNTER

This section is based upon a frequency counter you can construct in the lab. Figure 11-17 is a detailed diagram of that counter. This counter was designed using only parts you probably used in the lab. You will find this counter surprisingly accurate.

Figure 11-16 was the block diagram of the experimental frequency counter. Most parts in the wiring diagram in Fig. 11-17 are in the same general position as in the block diagram.

At the lower left in Fig. 11-17 is the 60-Hz sine wave being shaped into a square wave. This *wave shaping* is done by the 74121 one-shot multivibrator. This is the same unit we used in the digital clock to square up a sine wave. Remember that the following counter needs a square-wave input to operate properly.

To the right of the 74121 is a divide-by-6 counter. Three flip-flops (FF 1, FF 2, and FF 3) and a NAND gate are wired to form the modulo-6 counter. The frequency going into the divide-by-6 counter is 60 Hz; the frequency coming out of the counter (at Q of FF 3) is 10 Hz. The 10 Hz is fed back to the D flip-flop (FF 4). The D flip-flop (FF 4) just delays the pulse for $\frac{1}{60}$ s for timing purposes. The 10 Hz passes through FF 4 and toggles the J-K flip-flop (FF 5) to the count position ($\bar{Q} = 0$).

With the $\bar{Q}$ output of FF 5 at 0, the 74121 one-shot multivibrator at the upper left is activated. The one-shot multivibrator will let the unknown frequency through to the 10s counter. The top one-shot acts as a gate and also squares up the unknown input frequency. The 74121 IC will stay turned on by the *count pulse* for exactly $\frac{1}{10}$ s. During this time, pulses from the unknown frequency input are triggering the 10s counter, which counts the pulses. When the 10s counter goes from 9 to 10, it carries the 1 to the 100s counter. The contents of the counters are decoded and applied to the seven-segment displays. Output $\bar{Q}$ of FF 5 will finally toggle to 1. This turns off or inhibits the top 74121

one-shot, stopping the unknown frequency from getting to the 10s counter. So far only the *count* operation has been discussed. Remember that the frequency counter goes through a reset-count-display sequence.

The *display* part of the frequency counter's operation starts as the J-K flip-flop (FF 5) toggles, making output $\bar{Q}$ a logical 1. It was said that this 1 inhibits the 74121 IC. The 10s, 100s, and 1000s counters each hold a count in their flip-flops. This binary count is decoded by the three 7447 decoders. The decoders translate the input binary count to a seven-segment code. The proper segments of the seven-segment LEDs are lit, and the frequency appears on the display. For your convenience an extra zero (0) is added at the right of the display. This 0 must be added to the three seven-segment displays to have it read directly in hertz.

The display period on the frequency counter in Fig. 11-17 is about $\frac{9}{10}$ s. This fact may seem strange to you, considering the 10 Hz going into the J-K flip-flop (FF 5). You may have assumed that the next pulse from the divide-by-6 counter would toggle FF 5's output back to 0. This *would* happen, except the J and K inputs to FF 5 are driven to 0 by the output of the NOR gate (OR gate and inverter) at the right. This NOR gate holds the displays ON for $\frac{9}{10}$ s and then it goes to 1 for $\frac{1}{10}$ s. When the NOR puts out a 1, the pulse from the divide-by-6 counter toggles FF 5, which enables the 74121 IC. This 1 from the NOR gate lasts about $\frac{1}{10}$ s. This is just long enough for the count-up sequence to happen. Notice that just to the right of the NOR gate is a single inverter. This inverter produces the *display blanking* signal or pulse. When the NOR gate says count (1), the displays are temporarily *blanked* out by a logical 0 generated by the inverter.

The divide-by-10 counter at the bottom of Fig. 11-17 serves several purposes. The input to the divide-by-10 counter is 10 Hz; the output is 1 Hz. The four binary outputs of the 7493 counter are NORed. When all the outputs of the 7493 counter are 0, the NOR gate will generate the necessary 1 to enable FF 5 to toggle. This in turn enables the 74121 one shot. At all other times during the one-second cycle of the 7493 counter the output of the NOR gate is 0. This logical 0 disables FF 5 and therefore disables the 74121 one-shot multivibrator.

Experimental frequency counter

Display blanking

125

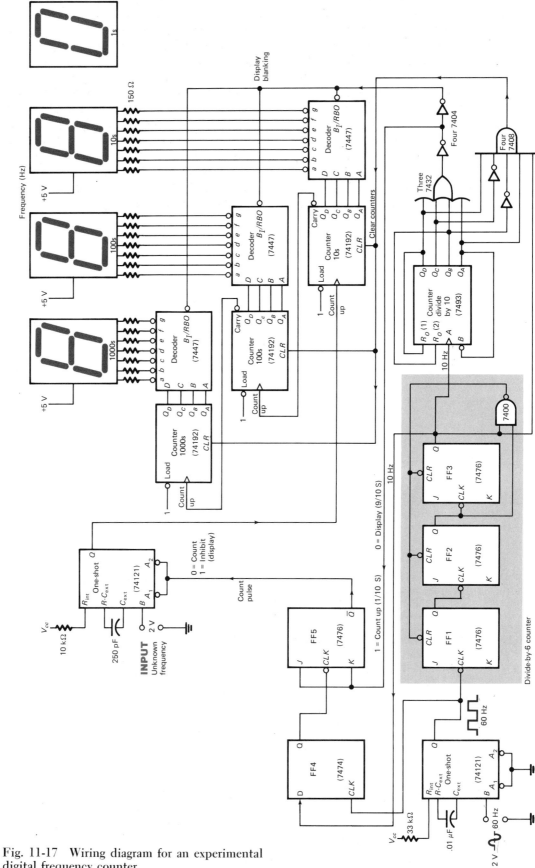

Fig. 11-17 Wiring diagram for an experimental
digital frequency counter.

A 5-input AND gate is shown at the lower right, Fig. 11-17. This AND gate generates a 1 for a short time just before the count-up sequence starts. This 1 from the AND gate *resets* or *clears* the counters (10s, 100s, 1000s) to 000. The output of the AND gate returns to its normal 0 during the count and display periods. The AND gate could be called the *counter reset* gate.

For the most part commercial frequency counters operate just like the one in Fig. 11-17. Commercial counters usually have more displays and read out in kilohertz and megahertz. The experimental frequency counter needed an input signal of about 2 V to make it operate. Commercial counters usually have an amplifier circuit before the first wave-shaping circuit to amplify weaker signals to the proper level. Overvoltage protection is also provided with a zener diode. To get rid of the blinking of the display, commercial counters usually use a slightly different method of storing and displaying the contents of the counters. We used the power line frequency of 60 Hz as our known frequency. Commercial frequency counters usually use an accurate high-frequency oscillator to generate their known frequency.

Summary

1. An assembly of subsystems connected correctly forms a digital system.
2. Systems have six common elements: input, transmission, storage, processing, control, and output.
3. Manufacturers produce ICs that are classed as small-, medium-, and large-scale integrations.
4. A calculator is a complex digital system often based upon a single LSI IC.
5. The computer is the most complex digital system. It is unique because of its vast size and stored program.
6. Multiplexers and demultiplexers are used to transmit parallel data through a single wire in serial form.
7. Errors occurring during data transmission can be detected by using parity bits.
8. A digital clock (watch) and a digital frequency counter are two closely related digital systems.
9. All digital systems are basically constructed from AND gates, OR gates, and inverters.

Questions

1. List at least five common devices that are considered digital systems.

2. List at least four devices you have used that are considered digital subsystems.

3. List the six elements found in most systems.

4. What do the following letters stand for when referring to ICs?
   a. IC   *c*. MSI
   b. SSI   *d*. LSI

5. The term *chip* usually is taken to mean a(n) ____?____ (IC, sliver of plastic) in digital electronics.

6. Inexpensive pocket calculators are usually based on a ____?____ (LSI, MSI).

7. The organization of the circuits within a calculator is called the ____?____ (architecture, dimensions) of the IC.

8. A calculator is *timed* by the ____?____ (internal clock, pressing of keys on the keyboard).

9. A simple calculator contains a rather large ____?____ (RAM, ROM).

10. ____?____ (Calculators, Computers, Both calculators and computers) contain a controller or control circuitry.

11. A ____?____ (computer, digital wristwatch) is usually based upon a single LSI IC.

12. Draw a diagram of the organization of the five main parts of a computer. Show the flow of *program* information and *data* through the system.

13. The CPU of a computer contains what three sections?

14. The ____?____ (ALU, CPU, MUX) section of a computer performs calculations and logic functions.

15. The most complex digital system is a ____?____ (computer, digital multimeter).

16. What do the following letters stand for?
    a. MUX
    b. DEMUX

17. A MUX-DEMUX system converts parallel input data to ____?____ (asynchronous, serial) data for transmission.

18. A MUX-DEMUX system operates somewhat like two ____?____ (rotary, three-way) switches.

19. Multiplexers are also known as ____?____ (data selectors, encoders). Demultiplexers are sometimes known as ____?____ (distributors, subtractors) or decoders.

20. It would take ____?____ (1, 5, 16) lines to transmit a 16-bit parallel word using a multiplexer-demultiplexer system.

21. Errors in transmission can be detected by using a ____?____ (parity, 16-word) bit.

22. A(n) ____?____ (AND, XOR) gate can detect an odd number of 1s at its input.

23. The ____?____ (Gray, Hamming) code is an error-correcting code.

24. The adder/subtractor in Fig. 11-10 contains which of the elements of a system listed in Fig. 11-1?

25. A digital clock system is closely related to a ____?____ (computer, frequency counter) system.

26. A digital clock makes extensive use of ____?____ (counter, shift register) subsystems.

27. A known frequency is the main input to a digital ____?____ (clock, frequency counter) system.

28. Counters are used for counting upward and ____?____ (shifting data, storing data) in the digital clock system.

29. Counters are used for counting upward and ____?____ (counting downward, dividing frequency) in a digital frequency counter.

30. Commercial digital clocks are usually constructed using ____?____ (LSIs, MSIs).

31. The one-shot multivibrator was used in the digital clock to ____?____ (count upward, square up the input wave).

32. The top 74121 one-shot multivibrator in Fig. 11-17 served as a ____?____ (storage register, wave shaper) and as a ____?____ (gate, shift register).

33. The three J-K flip-flops (FF 1, FF 2, FF 3) and the NAND gate in Fig. 11-17 function as a ____?____ (down counter, frequency divider).

34. The 7408 AND gate in Fig. 11-17 serves to ____?____ (clear, inhibit) the counters.

35. The frequency counter in Fig. 11-17 will count from a low of ____?____ Hz to a high of ____?____ Hz.

Connecting with Analog Devices

- To this point, all information entering or leaving a digital system has been digital information. Both inputs and output were typically in decimal or binary. Some digital systems, however, have *analog* inputs that vary *continuously* between two voltage levels. In this chapter we shall discuss the interfacing of analog devices.

The digital system in Fig. 12-1 has an analog input. The voltage varies continuously from 0 to 3 V. The *encoder* is a special device that converts the analog signal into digital information. We call this encoder an *analog-to-digital converter*, or an *A/D converter* for short. The A/D converter, then, converts analog information into digital information.

The digital system diagramed in Fig. 12-1 also has a *decoder*. This decoder is of a special type: it converts the digital information from the digital processing unit to an analog output. For instance, the analog output may be a continuous voltage change from 0 to 3 V. We call this decoder a *digital-to-analog converter*, or a *D/A converter* for short. The D/A converter, then, decodes digital information into an analog form.

The entire system in Fig. 12-1 might be called a *hybrid* system because it contains both digital and analog devices. The encoders and decoders that convert from analog-to-digital-to analog are called *interface* devices by engineers. The word "interface" is generally used when referring to a device or circuit that converts from one mode of operation to another. In this case we are converting from analog to digital data.

12-1 D/A CONVERSION

Refer to the D/A converter in Fig. 12-1. Let us suppose we want to convert the binary from the processing unit into a 0 to 3-V output. As with any decoder we must first set up a truth table of all the possible situations. Table 12-1 shows four inputs (D, C, B, A) into the D/A converter. The inputs are in binary form. Each 1 is about +3 to 5 V. Each 0 is about 0 V. The outputs are shown as voltages in the right column in Table 12-1. According to the table, if binary 0000 appears at the input of the D/A converter, the output is 0 V. If binary 0001 is the input, the output is 0.2 V. If binary 0010 appears at the input, then the output is 0.4 V. Notice that for each row you progress downward in Table 12-1, the analog output increases by 0.2 V.

A block diagram of a D/A converter is shown in Fig. 12-2. The digital inputs (D, C, B, A) are at the left. The decoder consists of two sections: the *resistor network* and the *summing amplifier*. The output is shown as a voltage reading on the voltmeter at the right.

The resistor network in Fig. 12-2 must take into account that a 1 at input B will be worth twice as much as a 1 at input A. Also, input C

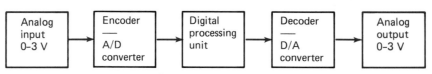

Fig. 12-1   A digital system with analog input and analog output.

130

Table 12-1   Truth table for D/A converter.

	DIGITAL INPUT				ANALOG OUTPUT
	D	C	B	A	Volts
Row 1	0	0	0	0	0
Row 2	0	0	0	1	0.2
Row 3	0	0	1	0	0.4
Row 4	0	0	1	1	0.6
Row 5	0	1	0	0	0.8
Row 6	0	1	0	1	1.0
Row 7	0	1	1	0	1.2
Row 8	0	1	1	1	1.4
Row 9	1	0	0	0	1.6
Row 10	1	0	0	1	1.8
Row 11	1	0	1	0	2.0
Row 12	1	0	1	1	2.2
Row 13	1	1	0	0	2.4
Row 14	1	1	0	1	2.6
Row 15	1	1	1	0	2.8
Row 16	1	1	1	1	3.0

ing a resistive ladder network, and an op amp used as the summing amplifier.

12-2 OPERATIONAL AMPLIFIER

The special amplifiers called op amps are characterized by high input impedance, low output impedance, and a variable voltage gain that can be set with external resistors. The symbol for an op amp is shown in Fig. 12-3(a). The op amp shown has two inputs. The top input is labeled as an inverting input. The inverting input is shown by the minus sign ($-$) on the symbol. The other input is labeled as a noninverting input. The noninverting input is shown by the plus sign ($+$) on the symbol. The output of the amplifier is also shown on the right of the symbol.

The operational amplifier is almost never used alone. Typically, the two resistors shown in Fig. 12-3(b) are added to the op amp to set the voltage gain of the amplifier. Resistor R_{in} is called the input resistor. Resistor R_f is called the feedback resistor. The *voltage gain* of this amplifier is found by using the simple formula

$$A_v \text{ (voltage gain)} = \frac{R_f}{R_{in}}$$

Suppose the value of the resistors connected to the op amp are $R_f = 10 \text{ k}\Omega$ and $R_{in} = 10 \text{ k}\Omega$. Using our voltage-gain formula, we find that

$$A_v = \frac{R_f}{R_{in}} = \frac{10,000}{10,000} = 1$$

will be worth four times as much as a 1 at input A. Several arrangements of resistors are used to do this job. These circuits are called *resistive ladder networks*.

The summing amplifier in Fig. 12-2 takes the output voltage from the resistor network and amplifies it the proper amount to get the voltages shown in the right column of Table 12-1. The summing amplifier typically uses an IC unit called an *operational amplifier*. An *op*erational *amp*lifier is often simply called an *op amp*.

The special decoder called a D/A converter consists of two parts: a group of resistors form-

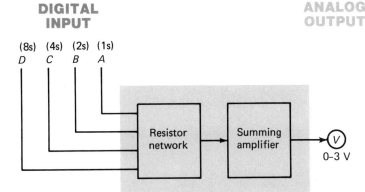

DIGITAL
INPUT

ANALOG
OUTPUT

(8s) (4s) (2s) (1s)
 D   C   B   A

Resistor
network

Summing
amplifier

0–3 V

D/A converter

Fig. 12-2   Block diagram of a D/A converter.

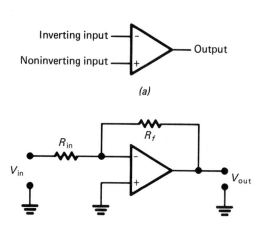

(a)

(b)

Fig. 12-3 Operational amplifier. (a) Symbol.
(b) With input and feedback resistors for setting
gain.

The gain of the amplifier will be 1. In our example, if the input voltage at V_{in} in Fig.
12-3(b) is 5 V, the output voltage at V_{out} will
be 5 V. The inverting input is being used, so
if the input voltage (V_{in}) is +5 V, then the
output voltage (V_{out}) will be −5 V. The voltage gain (A_v) of the op amp can also be calculated using the formula

$$A_v = \frac{V_{out}}{V_{in}}$$

The voltage gain for the circuit above will
then be

$$A_v = \frac{V_{out}}{V_{in}} = \frac{5}{5} = 1$$

The voltage gain is found to be 1 also using
this formula.

Suppose the input and feedback resistors
are 1 kΩ and 10 kΩ, as shown in Fig. 12-4.
What is the voltage gain for this circuit? The
voltage gain is calculated as

$$A_v = \frac{R_f}{R_{in}} = \frac{10,000}{1000} = 10$$

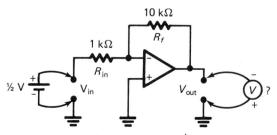

Fig. 12-4 Amplifier circuit using an op amp.

The voltage gain (A_v) was 10. If the input
voltage (V_{in}) is +0.5 V, then the voltage at the
output (V_{out}) will be how many volts? If the
gain is 10, then the input voltage of 0.5 V
times 10 equals 5 V. The output voltage at
V_{out} will be −5 V, as measured on the voltmeter in Fig. 12-4.

You have seen how the voltage gain (A_v) of
an op amp can be changed by changing the
ratio between the input and feedback resistors. You should know how to set the gain of
an operational amplifier by using different
values of resistors for R_{in} and R_f.

In summary, the op amp is part of a D/A
converter; it is used as a summing amplifier in
the converter. The gain of the op amp is easily set by the ratio of the input and feedback
resistors.

12-3 A BASIC D/A CONVERTER

A simple D/A converter is shown in Fig. 12-5.
The D/A converter is made in two sections.
The resistor network on the left is made up of
resistors R_1, R_2, R_3, and R_4. The summing
amplifier on the right consists of an op amp
and a feedback resistor. The input (V_{in}) is
3 V applied to switches D, C, B, and A. The
output voltage (V_{out}) is measured on a voltmeter. Notice that the op amp requires a
rather unusual dual power supply: a +10-V
power supply and a −10-V supply.

With all switches open, as shown in Fig.
12-5, the input voltage at point A will be 0 V
and the output voltage (V_{out}) will be 0 V.
This corresponds to row 1, Table 12-1. Suppose we close switch A in Fig. 12-5. The
input voltage of 3 V is applied to the op amp.
We next calculate the gain of the amplifier.
The gain is dependent upon the feedback resistor (R_f), which is 10 kΩ, and the input resistor (R_{in}), which is the value of R_1 or 150 kΩ.
Using the gain formula, we have

$$A_v = \frac{R_f}{R_{in}} = \frac{10,000}{150,000} = 0.066$$

To calculate the output voltage (V_{out}), we
multiply the gain by the input voltage as shown
here:

$$V_{out} = A_v \times V_{in} = 0.066 \times 3 = 0.2 \text{ V}$$

The output voltage (V_{out}) is 0.2 V when the

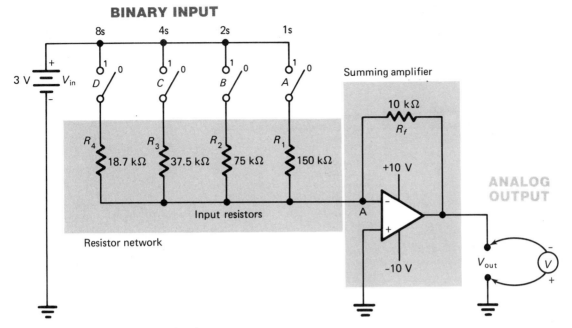

Fig. 12-5   A D/A converter circuit.

input is a binary 0001.   This satisfies the requirements of row 2, Table 12-1.

Let us now apply binary 0010 to the D/A converter in Fig. 12-5.   Switch B is closed, applying 3 V to the op amp.   The gain is

$$A_v = \frac{R_f}{R_{in}} = \frac{10,000}{75,000} = 0.133$$

Multiplying the gain times the input voltage will give us 0.4 V.   The 0.4 V is the output voltage (V_{out}).   This satisfies row 3, Table 12-1.

Notice that for each binary count on Table 12-1 the output voltage of the D/A converter increases by 0.2 V.   This increase occurs because of the increased voltage gain of the op amp as we switch in different resistors (R_1, R_2, R_3, R_4).   If only resistor R_4 from Fig. 12-5 were connected by closing switch D, the gain would be

$$A_v = \frac{R_f}{R_{in}} = \frac{10,000}{18,700} = 0.535$$

The gain multiplied by the input voltage of 3 V gives 1.6 V at the output of the op amp. This is what is required by row 9, Table 12-1.

For each increase in the binary count in Table 12-1 we have an increase of 0.2 V at the output of the D/A converter.   When all switches are closed (at logical 1) in Fig. 12-5,

the op amp puts out the full 3 V because the gain of the amplifier has increased to 1.

Any input voltage (V_{in}) up to the limits of the operational amplifier power supply (±10 V) may be used.   More binary places may be added by adding switches.   If a 16s place-value switch were added in Fig. 12-5, it would need a resistor with half the value of resistor R_4.   Its value would then have to be 9350 Ω.   The value of the feedback resistor (R_f) would also be changed to about 5 kΩ. The input would then contain a 5-bit binary number; the output would still be an analog output varying from 0 to 3 V.

The basic D/A converter shown in Fig. 12-5 does have two disadvantages: it takes a large range of resistor values, and it has low accuracy.

12-4 LADDER-TYPE D/A CONVERTER

Digital-to-analog converters consist of a resistor network and a summing amplifier.   Figure 12-6 diagrams a second type of resistor network that will provide the proper weighting for the binary inputs.   This resistor network is sometimes called the R-2R *ladder* network. The advantage of this arrangement of resistors is that only two values of resistors are used. Resistors R_1, R_2, R_3, R_4, and R_5 are 20 kΩ

133

R-2R ladder
network

Transistor-
transistor logic
(TTL)

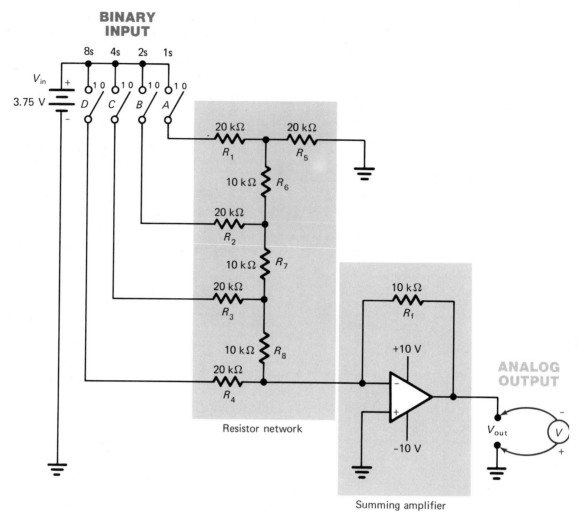

**BINARY
INPUT**

**ANALOG
OUTPUT**

Resistor network

Summing amplifier

Fig. 12-6   D/A converter circuit using an R-2R ladder resistor
networks.

each.   Resistors R_6, R_7, R_8, and R_f are each 10
kΩ.   Notice that all the horizontal resistors on
the "ladder" are exactly twice the value of the
vertical resistors, hence the title R-2R ladder
network.

The summing amplifier section in Fig. 12-6
is the same one used in the last unit.   Again
notice the use of the dual power supply on the
op amp.

The operation of this D/A converter is similar to the basic one in the last unit.   Table
12-2 details the operation of this D/A converter.   Notice that we are using an input
voltage of 3.75 V on this converter.   Each binary count increases the analog output by
0.25 V, as shown in the right column of Table
12-2.   Remember that each 0 on the input
side of the table means 0 V applied to that
input.   Each 1 on the input side of the table
means 3.75 V is applied to that input.   The
input voltage (V_{in}) of 3.75 V was used because

this is very close to the output of transistor-
transistor logic (TTL) counters and other ICs
you may have used.   The inputs (D, C, B, A)
in Fig. 12-6, then, could be connected directly
to the outputs of a TTL integrated circuit and
operate according to Table 12-2.   But in actual practice the outputs of a TTL integrated
circuit would not be accurate enough; they
would have to be put through a level translator to get a very precise voltage output.   More
binary places (16s, 32s, 64s, and so on) can be
added to the D/A converter in Fig. 12-6.   Follow the same pattern of resistor values as
shown in this diagram when adding place
values.

Two types of special decoders called digital-
to-analog converters have been covered.   The
R-2R ladder-type D/A converter has some advantages over the more basic unit.   The heart
of the D/A converter consists of the resistor
network and the summing amplifier.

Table 12-2   Truth table for D/A converter.

BINARY INPUT				ANALOG OUTPUT
8s	4s	2s	1s	
D	C	B	A	Volts
0	0	0	0	0
0	0	0	1	0.25
0	0	1	0	0.50
0	0	1	1	0.75
0	1	0	0	1.00
0	1	0	1	1.25
0	1	1	0	1.50
0	1	1	1	1.75
1	0	0	0	2.00
1	0	0	1	2.25
1	0	1	0	2.50
1	0	1	1	2.75
1	1	0	0	3.00
1	1	0	1	3.25
1	1	1	0	3.50
1	1	1	1	3.75

SELF TEST

Check your understanding by answering questions 1 to 10.

1. A special encoder that converts from analog to digital information is called a(n) ____?____ ____?____.

2. A special decoder that converts from digital to analog information is called a(n) ____?____ ____?____.

3. A D/A converter consists of a ____?____ network and a(n) ____?____ amplifier.

4. The name op amp stands for ____?____ ____?____.

5. What is the voltage gain (A_v) of an op amp such as the one shown in Fig. 12-3(b) if $R_{in} = 1$ kΩ and $R_f = 20$ kΩ?

6. What is the output voltage (V_{out}) from the op amp in question 5 if the input voltage (V_{in}) is $+0.2$ V?

7. Calculate the voltage gain of the op amp in Fig. 12-5 when only switch C (the 4s switch) is closed (at logical 1).

8. Using the voltage gain from question 7, calculate the output voltage of the D/A converter in Fig. 12-5 when only switch C is closed (at logical 1).

9. Refer to Fig. 12-6. The gain of the op amp is the greatest when all input switches are ____?____ (closed, open).

10. Refer to Fig. 12-6 and Table 12-2. The gain of the op amp is the *least* when switch ____?____ (A, B, C, D) is the only switch closed (at logical 1).

12-5 AN A/D CONVERTER

An analog-to-digital converter is a special type of encoder. A basic block diagram of an A/D converter is shown in Fig. 12-7. The input is a single-variable voltage. The voltage in this case will vary from 0 to 3 V. The output of the A/D converter will be in binary. The A/D converter will translate the analog voltage at the input into a 4-bit binary word. As with

Analog-to-digital converter

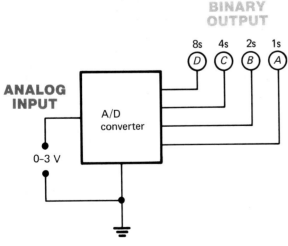

Fig. 12-7   Block diagram of an A/D converter.

Comparator

AND gate

BCD counter

D/A converter

Table 12-3 Truth table for A/D converter.

	ANALOG INPUT	BINARY OUTPUT			
	Volts	8s D	4s C	2s B	1s A
Row 1	0	0	0	0	0
Row 2	0.2	0	0	0	1
Row 3	0.4	0	0	1	0
Row 4	0.6	0	0	1	1
Row 5	0.8	0	1	0	0
Row 6	1.0	0	1	0	1
Row 7	1.2	0	1	1	0
Row 8	1.4	0	1	1	1
Row 9	1.6	1	0	0	0
Row 10	1.8	1	0	0	1
Row 11	2.0	1	0	1	0
Row 12	2.2	1	0	1	1
Row 13	2.4	1	1	0	0
Row 14	2.6	1	1	0	1
Row 15	2.8	1	1	1	0
Row 16	3.0	1	1	1	1

other encoders, it is well to define exactly the expected inputs and outputs. The truth table in Table 12-3 shows how the A/D converter should work. Row 1 shows 0 V being applied to the input of the A/D converter. The out-

put is a binary 0000. Row 2 shows a 0.2-V input. The output is a binary 0001. Notice that each increase of 0.2 V increases the binary count by 1. Finally, row 16 shows that when the maximum of 3 V is applied to the input, the output reads a binary 1111. Notice that the truth table in Table 12-3 is just the reverse of the D/A converter truth table in Table 12-1; the inputs and outputs have just been reversed.

The truth table for the A/D converter looks quite simple. The electronic circuits that perform the task detailed in the truth table are somewhat more complicated. One type of A/D converter is diagramed in Fig. 12-8. The A/D converter contains a *comparator*, an AND gate, a BCD counter, and a D/A converter. All the sections of the A/D converter except the comparator are familiar to you.

The analog voltage is applied at the left of Fig. 12-8. The comparator checks the voltage coming from the D/A converter. If the analog input voltage at A is *greater than* the voltage at input B of the comparator, the clock is allowed to *increase* the count of the BCD counter. The count on the counter increases until the feedback voltage from the D/A converter becomes greater than the analog input voltage. At this point the comparator stops the counter from advancing to a higher count. Suppose the input analog voltage

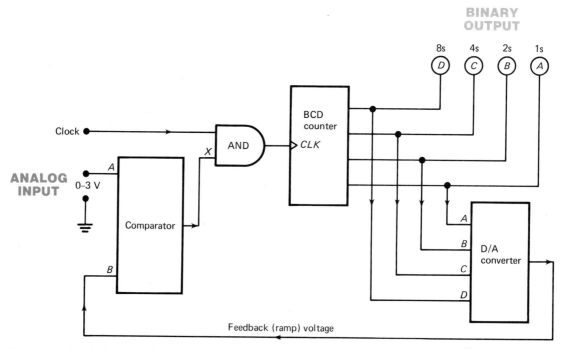

Fig. 12-8 Block diagram of a counter-ramp-type A/D converter.

136

were 2 V. According to truth table 12-3, the binary counter would have increased to 1010 before it was stopped. The counter is reset to binary 0000, and the counter starts counting again.

Now for more detail on the A/D converter in Fig. 12-8. Let us assume that there is a logical 1 at point X at the output of the comparator. Also assume that the BCD counter is at binary 0000. Assume too that 0.55 V is applied to the analog input. The 1 at point X enables the AND gate, and the first pulse from the clock appears at the CLK input of the BCD counter. The counter advances its count to 0001. The 0001 is displayed on the lights in the upper right of Fig. 12-8. The 0001 is also fed back to the D/A converter.

According to Table 12-1, a binary 0001 will produce a 0.2 V at the output of the D/A converter. The 0.2 V is fed back to the B input of the comparator. The comparator checks its inputs. The A input is higher (0.55 V as opposed to 0.2 V), so the comparator puts out a logical 1. The 1 enables the AND gate, which lets the next clock pulse through to the counter. The counter advances its count by one. The count is now 0010. The 0010 is fed back to the D/A converter.

According to Table 12-1, a 0010 input will produce a 0.4-V output. The 0.4 V is fed back to the B input of the comparator. The comparator again checks the B input against the A input; the A input is still larger (0.55 V as opposed to 0.4 V). The comparator puts out a logical 1. The AND gate is enabled, letting the next clock pulse reach the counter. The counter increases its count to a binary 0011. The 0011 is fed back to the D/A converter.

According to Table 12-1, the output will be 0.6 V. The 0.6 V is fed back to the B input of the comparator. The comparator checks input A against input B; for the first time the B input is larger than the A input. The comparator puts out a logical 0. The logical 0 will disable the AND gate. No more clock pulses can reach the counter. The counter has stopped at binary 0011. Binary 0011 must then equal 0.55 V. A look at row 4, Table 12-3, shows that 0.6 V gives the readout of binary 0011. Our A/D converter has worked according to the truth table.

If the input analog voltage were 1.2 V, the binary output would be 0110, according to Table 12-3. The counter would have to count from binary 0000 to 0110 before being stopped by the comparator. If the input analog voltage were 2.8 V, the binary output would be 1110. The counter would have to count from binary 0000 to 1110 before being stopped by the comparator. Notice that it does take some time for the conversion of the analog voltage to a binary readout. However, in most cases the clock runs fast enough so that this time lag is not a problem.

You now should appreciate why we studied the D/A converter before the A/D converter. This *counter-ramp A/D converter* is fairly complex and needs a D/A converter to operate. The term *ramp* in the title for this converter refers to the gradually increasing voltage from the D/A converter that is fed back to the comparator. If you drew a graph of the voltage being fed to input B of the comparator, it would appear as a *ramp* or a *sawtooth waveform*.

12-6 COMPARATORS

In the last section we used a *comparator*. We found that a comparator will compare two voltages and tell us which is the larger of the two. Figure 12-9 is a basic block diagram of a comparator. If the voltage at input A is larger than at input B, the comparator gives a 1 output. If the voltage at input B is larger than at input A, the output is a logical 0. This is written as $A > B = 1$ or $B > A = 0$ in Fig. 12-9.

The heart of a comparator is an op amp. Figure 12-10(a) shows a comparator circuit. Notice that input A has 1.5 V applied and input B has 0 V applied. We would find that the output voltmeter would read about 3.5 V or a logical 1.

Figure 12-10(b) shows that the input B voltage has been increased to 2 V. Input A is still at 1.5 V. Input B is larger than input A. The output of the comparator circuit will be about 0 V (actually the voltage will be about -0.6 V) or a logical 0.

The comparator in the A/D converter in

Counter-ramp A/D converter

Ramp voltage

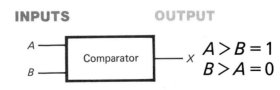

Fig. 12-9 Block diagram of a comparator.

Zener diode

Digital voltmeter

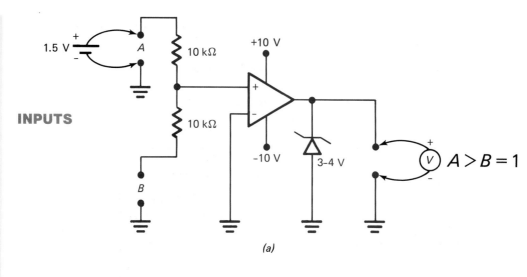

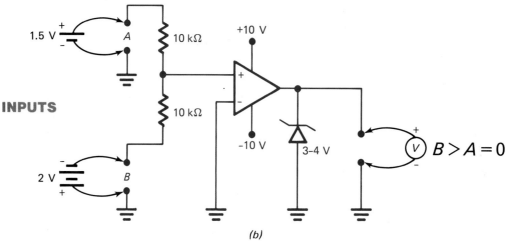

Fig. 12-10   Comparator circuit.   (a) With greater voltage at input A.   (b) With greater voltage at input B.

Fig. 12-8 works exactly like this unit. The zener diode in the comparator in Fig. 12-10 is there to clamp the output voltage at about +3.5 and 0 V. Without the zener diode the output voltages would be about +9 and −9 V. The +3.5 and 0 V are more compatible with the TTL integrated circuits you may have used.

12-7 A DIGITAL VOLTMETER

One use for an A/D converter is in a *digital voltmeter*. You have already used all the subsystems needed to make a digital voltmeter system. A block diagram of a simple digital voltmeter is shown in Fig. 12-11. The A/D

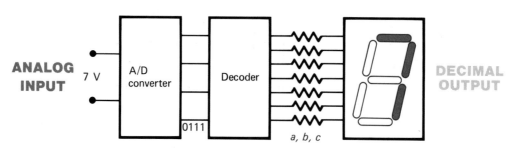

Fig. 12-11   Block diagram of a digital voltmeter.

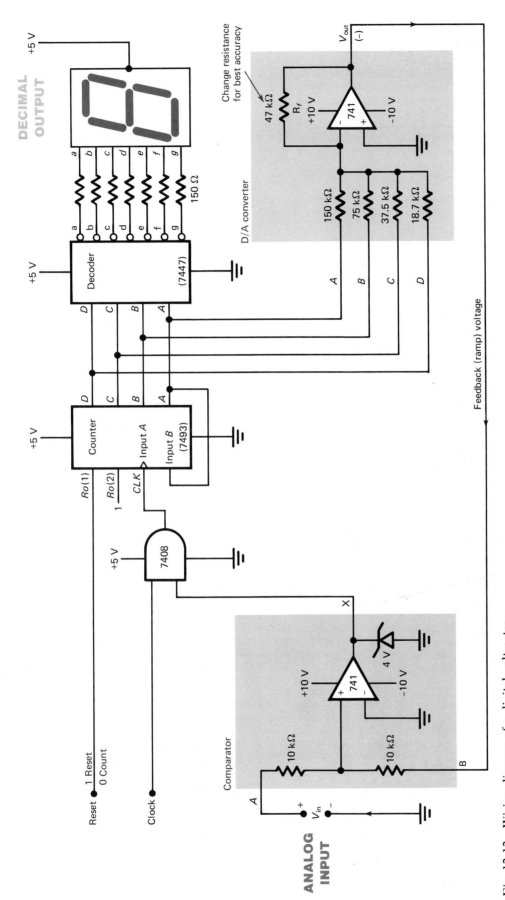

Fig. 12-12 Wiring diagram of a digital voltmeter.

Counter-ramp type A/D converter

Ramp-type A/D converter

Ramp generator

converter converts the analog voltage into a binary form. The binary is sent to the decoder, where it is converted to a seven-segment code. The seven-segment readout indicates the voltage in decimals. With 7 V applied to the input of the A/D converter, the unit will put out a binary 0111, as shown. The decoder activates lines *a* to *c* of the seven-segment display; segments *a* to *c* light on the display. The display reads as a decimal 7. Note that the A/D converter is also a decoder; it decodes from an analog input to a binary output.

A wiring diagram of a digital voltmeter is shown in Fig. 12-12. Notice the comparator, the AND gate, the counter, the decoder, the seven-segment display, and the D/A converter. Several power supplies are needed to set up this circuit. A dual ±10-V supply (or two individual 10-V supplies) is used for the 741 op amps. A 5-V supply is used for the 7408, 7493, and 7447 ICs, and the seven-segment LED display. A 0- to 10-V variable direct current (dc) power supply is also needed for the analog input voltage.

Let us assume a 2-V input to the analog input of the digital voltmeter in Fig. 12-12. Reset the counter to 0000. The comparator checks inputs A and B; A is larger (A = 2 V, B = 0 V). The comparator output is a logical 1. This 1 enables the AND gate. The pulse from the clock passes through the AND gate. The pulse causes the counter to advance one count. The count is now 0001. The 0001 is applied to the decoder. The decoder enables lines *b* and *c* to the seven-segment display; segments *b* and *c* light on the display, giving a decimal readout of 1. The 0001 is also applied to the D/A converter. About 3.2 V from the counter is applied through the 150 kΩ resistor to the input of the op amp. The voltage gain of the op amp is

$$A_v = \frac{R_f}{R_{in}} = \frac{47,000}{150,000} = 0.31$$

The gain is 0.31. The voltage gain times the input voltage equals the output voltage:

$$V_{out} = A_v \times V_{in} = 0.31 \times 3.2 = 1 \text{ V}$$

The output voltage of the D/A converter is −1 V. The 1 V is fed back to the comparator.

Now, with the 2 V still applied to the input,

the comparator checks A against B; input A is larger. The comparator applies a logical 1 to the AND gate. The AND gate passes the second clock pulse to the counter. The counter advances to 0010. The 0010 is decoded and reads out as a decimal 2 on the seven-segment display. The 0010 also is applied to the D/A converter. The D/A converter puts out about 2 V, which is fed back to the B input of the comparator.

The display now reads 2. The 2 V is still applied to input A of the comparator. The comparator checks A against B; B is just slightly larger. Output X of the comparator goes to logical 0. The AND gate is disabled. No clock pulses reach the counter. The count has stopped at 2 on the display. This is the voltage applied at the analog input.

It is highly suggested that you set up a D/A converter, a comparator, and an A/D converter (such as the digital voltmeter). The details of operation of these interface devices will be appreciated only by your wiring and using them in the lab.

12-8 OTHER A/D CONVERTERS

In Sec. 12-5 we studied the counter-ramp-type A/D converter. Several other types of A/D converters are also used; in this section we shall discuss two other types of converters.

A *ramp-type A/D converter* is shown in Fig. 12-13. This A/D converter works very similarly to the counter-ramp-type A/D converter in Fig. 12-8. The *ramp generator* at the left in Fig. 12-13 is the only new subsystem. The ramp generator produces a sawtooth waveform, which looks like the triangle-shaped wave in Fig. 10-14(*a*).

Suppose 3 V were applied to the analog voltage input of the A/D converter in Fig. 12-13. This situation is diagramed in Fig. 12-14(*a*). The ramp voltage starts to increase but is still lower than input A of the comparator. The comparator output is at a logical 1. This 1 enables the AND gate so a clock pulse can get through. In Fig. 12-14(*a*) the diagram shows three clock pulses getting through the AND gate before the ramp voltage gets larger than the input voltage. At point Y in Fig. 12-14(*a*) the comparator output goes to a logical 0. The AND gate is disabled. The counter stops counting at a binary 0011. The binary 0011 means 3 V is applied at the input.

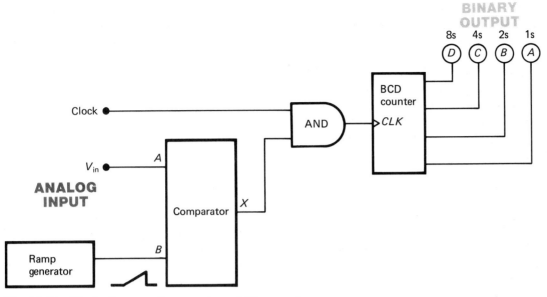

Fig. 12-13   Block diagram of a ramp-type A/D converter.

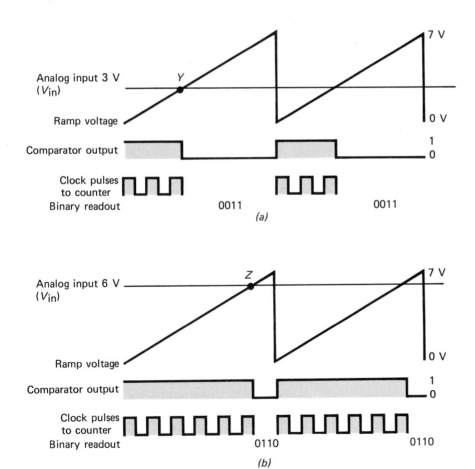

Fig. 12-14   Ramp-type A/D converter waveforms.   (*a*) With 3 V applied.   (*b*) With 6 V applied.

Successive-
approximation-
type A/D
converter

Figure 12-14(*b*) gives another example. The input voltage to the ramp-type A/D converter is 6 V in this situation. The ramp voltage begins to increase from left to right. The comparator output is at a logical 1 because input *A* is larger than the ramp generator voltage at input *B*. The counter continues to advance. At point Z on the ramp voltage the ramp generator voltage is larger than V_{in}. At this point the comparator output goes to a logical 0. This 0 disables the AND gate. The clock pulses no longer reach the counter. The counter is stopped at a binary 0110. The binary 0110 stands for the 6-V analog input.

The difficulty with the ramp-type A/D converters is the long time it takes to count up to higher voltages. For instance, if the binary output were eight binary places, the counter might have to count up to 255. To cure this slow conversion time we could use a different type of A/D converter. A converter that cuts down on conversion time is a *successive-approximation-type A/D converter*.

A block diagram of a successive-approximation-type A/D converter is shown in Fig. 12-15. The converter consists of a comparator, a D/A converter, and a new logic block.

The new logic block is called the successive-approximation logic section.

Suppose we apply 7 V to the analog input (V_{in}). The successive-approximation A/D converter first makes a "guess" at the analog input voltage. This guess is made by setting the most significant bit (MSB) to 1. This is shown in block 1, Fig. 12-16. This job is performed by the successive-approximation logic unit. The result (1000) is fed back to the comparator through the D/A converter. The comparator answers the question in block 2, Fig. 12-16, Is 1000 high or low compared with the input voltage? In this case the answer is high. The successive-approximation logic then performs the task in block 3. The 8s place is cleared to 0, and the 4s place is set to 1. The result (0100) is sent back to the comparator through the D/A converter. The comparator next answers the question in block 4. Is 0100 high or low compared with the input voltage? The answer is low. The successive-approximation logic then performs the task in block 5. The 2s place is set to 1. The result (0110) is sent back to the comparator. The comparator answers the question in block 6. Is 0110 high or low compared with

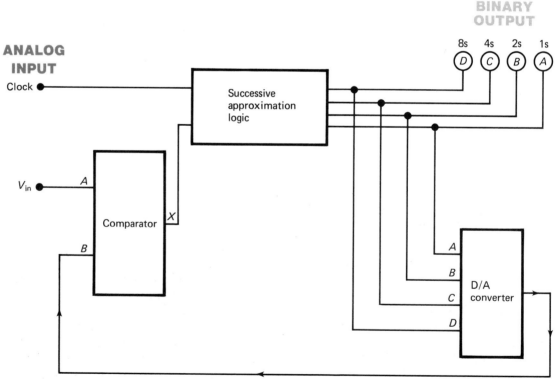

Fig. 12-15 Block diagram of a successive-approximation-type A/D converter.

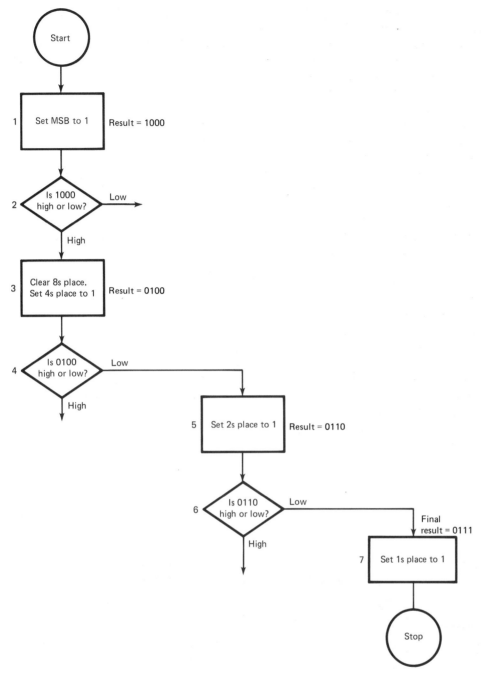

Digitizing
process

Fig. 12-16 Flowcharting the operation of the successive-approximation-type A/D converter.

the input voltage? The answer is low. The successive approximation logic then performs the task in block 7. The 1s place is set to 1. The final result is a binary 0111. This stands for the 7 V applied at the input of the A/D converter.

Notice in Fig. 12-16 that the items in the rectangles are performed by the successive-approximation logic unit. The questions are answered by the comparator. Also notice that the task performed by the successive-approximation logic depends upon whether the answer to the previous question was low or high (see blocks 3 and 5).

The advantage of the successive-approximation A/D converter is that it takes fewer guesses to get the answer. The *digitizing* process is thus faster. The successive-approximation-type A/D converter is very widely used.

143

Summary

1. Special interface encoders and decoders are used between analog and digital devices. These are called D/A converters and A/D converters.
2. A D/A converter consists of a resistor network and summing amplifier.
3. Operational amplifiers are used in D/A converters and comparators. Gain can be easily set with external resistors on the op amp.
4. Several different resistor networks are used for weighting the binary input to a D/A converter.
5. Common A/D converters are the counter-ramp, ramp-generator, and successive-approximation types.
6. A comparator compares two voltages and determines which is larger. An operational amplifier is the heart of the comparator.
7. A digital voltmeter is one application of an A/D converter.

Questions

1. An analog-to-digital converter is a special type of ____?____ (decoder, encoder).

2. A D/A converter is a(n) ____?____ (decoder, encoder).

3. The ____?____ (A/D, D/A) converter will digitize information.

4. The ____?____ (A/D, D/A) converter will translate from binary to an analog voltage.

5. A D/A converter consists of a ____?____ network and a summing ____?____ .

6. Operational amplifier is frequently shortened to ____?____ ____?____ .

7. The voltage gain of the operational amplifier in Fig. 12-3(b) is determined by dividing the value of ____?____ (R_f, R_{in}) by the value of ____?____ (R_f, R_{in}).

8. Draw a symbol for an operational amplifier. Label the inverting input with a $-$ and the noninverting input with a $+$. Label the output. Label the $+10$-V and -10-V power-supply connections.

9. Refer to Fig. 12-4. What is the gain (A_v) of the op amp in this diagram if $R_{in} = 1$ kΩ and $R_f = 100$ kΩ?

10. Refer to Fig. 12-4. With the input voltage as $+\frac{1}{2}$ V, the output voltage will be ____?____ ($+$, $-$) 5 V. This is because we are using the ____?____ (inverting, noninverting) input.

11. Refer to Fig. 12-5. What is the voltage gain of the op amp in this circuit with only switch A closed (at logical 1)?

12. Refer to Fig. 12-5. What is the combined resistance of parallel resistors R_1 and R_2 if both switches A and B are closed (at logical 1)?

13. Refer to Fig. 12-5. What is the gain (A_v) of the op amp with switches A and B closed (at logical 1)? (Use the resistance figure from question 12.)

14. Refer to Fig. 12-5. What is the output voltage (V_{out}) when a binary 0011 is applied to the D/A converter? (Use the A_v from question 13.)

15. The arrangement of resistors in Fig. 12-6 is called the ____?____ ladder network.

16. Compare Tables 12-1 and 12-2. The difference between the two would be the ___?___ (binary inputs, scaling of the analog output).

17. A high or logical 1 from a TTL device is about ___?___ (0, 3.75, 5.5) V.

18. The letters TTL stand for ___?___ ___?___ ___?___ .

19. The ___?___ (A/D, D/A) converter is the more complicated electronic system.

20. The counter-ramp-type A/D converter consists of an ___?___ (AND, OR) gate, a ___?___ (comparator, resistor network), a ___?___ (counter, shift register), and a D/A ___?___ .

21. Refer to Fig. 12-8. If point X is at a logical ___?___ (0, 1), the counter will advance a count as a pulse comes from the clock.

22. Refer to Fig. 12-8. If input B of the comparator has a higher voltage than input A, the AND gate will be ___?___ (disabled, enabled).

23. Refer to Fig. 12-9. If input A of the comparator equals 5 V and input B equals 2 V, then output X will be a logical ___?___ (0, 1). Output X would be about ___?___ (0, 4) V.

24. The primary component in a comparator is a(n) ___?___ (counter, op amp).

25. Refer to Fig. 12-12. This digital voltmeter uses a ___?___ (counter-ramp, successive-approximation)-type A/D converter.

26. A ramp-type A/D converter consists of an ___?___ (AND, OR) gate, a ___?___ (counter, register), a ramp ___?___ , and a ___?___ (comparator, D/A converter).

27. The ___?___ (ramp, successive-approximation)-type A/D converter is faster at digitizing information.

Answers to Self Test

1. A/D converter
2. D/A converter
3. Resistor, summing (scaling)
4. Operational amplifier
5. $A_v = 20$
6. $V_{out} = -4$ V
7. $A_v = 0.266$
8. $V_{out} = -0.8$ V
9. Closed
10. A

Index